D0557712

# Big Data and the Internet of Things

## Enterprise Information Architecture for a New Age

Robert Stackowiak

Art Licht

Venu Mantha

Louis Nagode

Apress®

ISBN-13 (pbk): 978-1-4842-0987-5

ISBN-13 (electronic): 978-1-4842-0986-8

Managing Director: Welmoed Spahr
Lead Editor: Jonathan Gennick
Editorial Board: Steve Anglin, Mark Beckner, Gary Cornell, Louise Corrigan, Jim DeWolf,
    Jonathan Gennick, Robert Hutchinson, Michelle Lowman, James Markham,
    Matthew Moodie, Jeffrey Pepper, Douglas Pundick, Ben Renow-Clarke, Gwenan Spearing,
    Matt Wade, Steve Weiss
Coordinating Editor: Jill Balzano
Copy Editor: Ann Dickson
Compositor: SPi Global
Indexer: SPi Global
Artist: SPi Global
Cover Designer: Anna Ishchenko

Distributed to the book trade worldwide by Springer Science+Business Media New York, 233 Spring Street, 6th Floor, New York, NY 10013. Phone 1-800-SPRINGER, fax (201) 348-4505, e-mail orders-ny@springer-sbm.com, or visit www.springeronline.com. Apress Media, LLC is a California LLC and the sole member (owner) is Springer Science + Business Media Finance Inc (SSBM Finance Inc). SSBM Finance Inc is a Delaware corporation.

For information on translations, please e-mail rights@apress.com, or visit www.apress.com.

Apress and friends of ED books may be purchased in bulk for academic, corporate, or promotional use. eBook versions and licenses are also available for most titles. For more information, reference our Special Bulk Sales–eBook Licensing web page at www.apress.com/bulk-sales.

Any source code or other supplementary material referenced by the author in this text is available to readers at www.apress.com. For detailed information about how to locate your book's source code, go to www.apress.com/source-code/.

*This book is dedicated to pioneers for whom technology provides a means to a business solution.*

## Praise

*This book is an absolute must-read for any business or technical professional today who is tasked with how to architect and deliver on today's complex enterprise informational needs. Bob and his team clearly articulate how to address those needs in a clear and concise manner, and take you through the process of turning the Big Data and Signal Data into a powerful enterprise asset.*

—Richard J. Solari, Director, Information Management, Deloitte

*This book is a great starting point for enterprise architects who need to establish a reference architecture and roadmap for Big Data and Internet of Things implementations. Using an approach familiar to EA practitioners, and by providing valuable insights into business motivation, the technologies and their implications, the authors guide readers through the process of building a complete, relevant, and coherent story tuned to the specific needs of their organizations.*

—George S. Paras, Managing Director, EAdirections

*This book provides a wealth of information on defining and developing a project in which technical infrastructure evolves from traditional data management platforms to an infrastructure that also includes Hadoop and NoSQL databases. It is one of few books that I have seen that addresses how to define and develop solutions that must be deployed on modern information management architecture.*

—Paul Cross, SVP Co-Prime Sales, Salesforce.com

# Contents at a Glance

# Contents

# About the Authors

**Robert Stackowiak** is Vice President of Information Architecture and Big Data at Oracle in North America. His team of architects and experts focuses on Big Data (including Hadoop and NoSQL databases), predictive analytics, data warehousing, business intelligence, and information discovery. The team engages with companies that are implementing these technologies and exploring new solutions such as those enabled by the Internet of Things. Bob has spoken at conferences around the world and co-authored many books on data management and business intelligence including five editions of *Oracle Essentials* (O'Reilly Media), *Oracle Big Data Handbook* (Oracle Press), *Achieving Extreme Performance with Oracle Exadata* (Oracle Press), and *Oracle Data Warehousing and Business Intelligence Solutions* (Wiley). Follow him on Twitter @rstackow.

**Art Licht** is a Senior Director and Distinguished Sales Consultant at Oracle. Focused on Information Architecture and Big Data, he has over 25 years of global and engineering experience in building and designing high-performance data management systems. Art has recently assisted companies in developing data management strategies that align to business priorities to enable growth, improve business agility, and reduce overall IT costs. He has published numerous best practices and technical white papers over the years. Prior to joining Oracle, Art was a Distinguished Engineer at Sun Microsystems.

**Venu Mantha** is a Senior Director of Information Architecture and Big Data with Oracle. He brings over 20 years of global management and technology advisory services experience. Venu has developed organizational growth strategies, led go-to-market initiatives, and spearheaded development of methods and tools for practitioner use. He has also led many technology initiatives to streamline and improve operations for global clients that involve transformations, cost reductions, consolidations, and post merger integrations. More recently, he has advised clients in multiple industries regarding data warehousing, analytics, Big Data, and the Internet of Things. Venu earned his MBA from the University of Michigan, Ross School of Business, with a focus on Corporate Strategy and General Management.

**Louis Nagode** is a Senior Director of Information Architecture and Big Data at Oracle. He has worked for over 30 years in business intelligence and in IT and development related roles including management of software and product development, sales and sales consulting, and business development. Louis has extensive experience performing proofs of concept at clients. As a by-product of this experience, he created Oracle's BI Challenge to Go (BIC2G), a portable environment used worldwide in demonstrations, workshops, and proofs of concept. His fun, plain-talking style makes him a sought after speaker. Louis has a bachelors degree and masters degree in Electrical Engineering and Computer Science from MIT.

# Acknowledgments

We begin our acknowledgements by first recognizing our clients. We are much smarter now than when we first considered writing this book three years ago. We led over two hundred workshops in that time using the methodology for success that we document in the book. Our clients are always looking to us to come up with new solutions to their business and technology challenges while considering their past investments and architecture designs. You will find that we followed an evolutionary path throughout the book based on this experience.

At Apress, we would like to thank Jonathan Gennick for editorial direction and support of this project. He first saw the value in publishing this work and shepherded it through the entire process. We also thank Jill Balzano, who served as coordinating editor on the project. Both Jonathan and Jill made writing this book a much easier process.

As all of us wrote this while at Oracle, we would like to acknowledge that our management was supportive of a book covering best practices that is entirely vendor independent. Among those supporting this effort were Joseph Strada and Anasuya Strasner. They and others we work with recognize that enterprise architecture solutions are rarely comprised of only one vendor's products or solutions.

During the development of information architecture workshops that formed the basis for some of the material in this book, many colleagues provided critiques and added content. Among those we'd like to thank are Jason Fish, Bob England, Alan Manewitz, Tom Luckenbach, Bob Cauthen, Linda McHale, and Khader Mohiuddin.

Given the significant role that enterprise architects have in defining these solutions, we would like to call attention to the strong partnership the authors have with the Oracle Enterprise Architecture community led by Hamidou Dia in North America and Andrew Bond in EMEA. Robert Stackowiak would also like to acknowledge a long friendship with George Paras of EAdirections whose guidance on best practices at a lunch long ago had influence on this book.

We also received a lot of input regarding the approach from friends at Accenture, Capgemini, Cloudera, Deloitte, mFrontiers, Onx, Optimal Design, Vlamis Software Solutions, and others. We validated the approach at various times with each of these organizations and sometimes engaged together using the approach at joint clients.

Given our roles, we also work closely with many in Oracle's product teams. The knowledge we gained over the years as to how these products are deployed in open architecture footprints was invaluable here. Among the product managers we would like to acknowledge are Neil Mendelson, George Lumpkin, Jean-Pierre Dijcks, Dan McClary, Marty Gubar, Hermann Baer, Maria Colgan, Mark Hornick, Ryan Stark, Richard Tomlinson, Jeff Pollock, and Harish Gaur.

Last, but certainly not least, writing a book is reserved for long plane rides and for times spent in the office on weekends and during late nights. We would like to acknowledge the support of our wives, Jodie Stackowiak, Shayne Licht, Geetha Mantha, and Jennifer Nagode. We couldn't do it without you.

# Introduction

The genesis of this book began in 2012. Hadoop was being explored in mainstream organizations, and we believed that information architecture was about to be transformed. For many years, business intelligence and analytics solutions had centered on the enterprise data warehouse and data marts, and on the best practices for defining, populating, and analyzing the data in them. Optimal relational database design for structured data and managing the database had become the focus of many of these efforts. However, we saw that focus was changing.

For the first time, streaming data sources were seen as potentially important in solving business problems. Attempts were made to explore such data experimentally in hope of finding hidden value. Unfortunately, many efforts were going nowhere. The authors were acutely aware of this as we were called into many organizations to provide advice.

We did find some organizations that were successful in analyzing the new data sources. When we took a step back, we saw a common pattern emerging that was leading to their success. Prior to starting Big Data initiatives, the organizations' stakeholders had developed theories about how the new data would improve business decisions. When building prototypes, they were able to prove or disprove these theories quickly.

This successful approach was not completely new. In fact, many used the same strategy when developing successful data warehouses, business intelligence, and advanced analytics solutions that became critical to running their businesses. We describe this phased approach as a methodology for success in this book. We walk through the phases of the methodology in each chapter and describe how they apply to Big Data and Internet of Things projects.

Back in 2012, we started to document the methodology and assemble artifacts that would prove useful when advising our clients, regardless of their technology footprint. We then worked with the Oracle Enterprise Architecture community, systems integrators, and our clients in testing and refining the approach.

At times, the approach led us to recommend traditional technology footprints. However, new data sources often introduced a need for Hadoop and NoSQL database solutions. Increasingly, we saw Internet of Things applications also driving new footprints. So, we let the data sources and business problems to be solved drive the architecture.

About two years into running our workshops, we noticed that though many books described the technical components behind Big Data and Internet of Things projects, they rarely touched on how to evaluate and recommend solutions aligned to the information architecture or business requirements in an organization. Fortunately, our friends at Apress saw a similar need for the book we had in mind.

This book does not replace the technical references you likely have on your bookshelf describing in detail the components that can be part of the future state information architecture. That is not the intent of this book. (We sometimes ask enterprise architects what components are relevant, and the number quickly grows into the hundreds.)

Our intent is to provide you with a solid grounding as to how and why the components should be brought together in your future state information architecture. We take you through a methodology that establishes a vision of that future footprint; gathers business requirements, data, and analysis requirements; assesses skills; determines information architecture changes needed; and defines a roadmap. Finally, we provide you with some guidance as to things to consider during the implementation.

We believe that this book will provide value to enterprise architects where much of the book's content is directed. But we also think that it will be a valuable resource for others in IT and the lines of business who seek success in these projects.

Helping you succeed is our primary goal. We hope that you find the book helps you reach your goals.

# CHAPTER 1

■ ■ ■

# Big Data Solutions and the Internet of Things

This book begins with a chapter title that contains two of the most hyped technology concepts in information architecture today: *Big Data* and the *Internet of Things*. Since this book is intended for enterprise architects and information architects, as well as anyone tasked with designing and building these solutions or concerned about the ultimate success of such projects, we will avoid the hype. Instead, we will provide a solid grounding on how to get these projects started and ultimately succeed in their delivery. To do that, we first review how and why these concepts emerged, what preceded them, and how they might fit into your emerging architecture.

The authors believe that Big Data and the Internet of Things are important evolutionary steps and are increasingly relevant when defining new information architecture projects. Obviously, you think the technologies that make up these solutions could have an important role to play in your organization's information architecture as you are reading this book. Because we believe these steps are evolutionary, we also believe that many of the lessons learned previously in developing and deploying information architecture projects can and should be applied in Big Data and Internet of Things projects.

Enterprise architects will continue to find value in applying agile methodologies and development processes that move the organization's vision forward and take into account business context, governance, and the evolution of the current state architecture into a desired future state. A critical milestone is the creation of a roadmap that lays out the prioritized project implementation phases that must take place for a project to succeed.

Organizations already successful in defining and building these next generation solutions have followed these best practices, building upon previous experience they had gained when they created and deployed earlier generations of information architecture. We will review some of these methodologies in this chapter.

On the other hand, organizations that have approached Big Data and the Internet of Things as unique technology initiatives, experiments, or resume building exercises often struggle finding value in such efforts and in the technology itself. Many never gain a connection to the business requirements within their company or organization. When such projects remain designated as purely technical research efforts, they usually reach a point where they are either deemed optional for future funding or declared outright failures. This is unfortunate, but it is not without precedence.

In this book, we consider Big Data initiatives that commonly include traditional data warehouses built with relational database management system (RDBMS) technology, Hadoop clusters, NoSQL databases, and other emerging data management solutions. We extend the description of initiatives driving the adoption of the extended information architecture to include the Internet of Things where sensors and devices with intelligent controllers are deployed. These sensors and devices are linked to the infrastructure to enable analysis of data that is gathered. Intelligent sensors and controllers on the devices are designed to trigger immediate actions when needed.

So, we begin this chapter by describing how Big Data and the Internet of Things became part of the long history of evolution in information processing and architecture. We start our description of this history at a time long before such initiatives were imagined. Figure 1-1 illustrates the timeline that we will quickly proceed through.

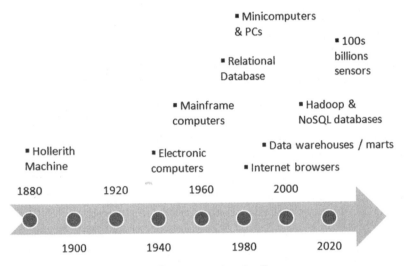

*Figure 1-1.* *Evolution in modern computing timeline*

# From Punched Cards to Decision Support

There are many opinions as to when modern computing began. Our historical description starts at a time when computing moved beyond mechanical calculators. We begin with the creation of data processing solutions focused on providing specific information. Many believe that an important early data processing solution that set the table for what was to follow was based on punched cards and equipment invented by Herman Hollerith.

The business problem this invention first addressed was tabulating and reporting on data collected during the US census. The concept of a census certainly wasn't new in the 1880s when Hollerith presented his solution. For many centuries, governments had manually collected data about how many people lived in their territories. Along the way, an expanding array of data items became desirable for collection such as citizen name, address, sex, age, household size, urban vs. rural address, place of birth, level of education, and more. The desire for more of these key performance indicators (KPIs) combined with population growth drove the need for a more automated approach to data collection and processing. Hollerith's punched card solution addressed these needs. By the 1930s, the technology had become widely popular for other kinds of data processing applications such as providing the footprint for accounting systems in large businesses.

The 1940s and the World War II introduced the need to solve complex military problems at a faster pace, including the deciphering of messages hidden by encryption and calculating the optimal trajectories for massive guns that fired shells. The need for rapid and incremental problem solving drove the development of early electronic computing devices consisting of switches, vacuum tubes, and wiring in racks that filled entire rooms. After the war, research in creating faster computers for military initiatives continued and the technology made its way into commercial businesses for financial accounting and other uses.

The following decades saw the introduction of modern software operating systems and programming languages (to make applications development easier and faster) and databases for rapid and simpler retrieval of data. Databases evolved from being hierarchical in nature to the more flexible relational model where data was stored in tables consisting of rows and columns. The tables were linked by foreign keys between common columns within them. The Structured Query Language (SQL) soon became the standard means of accessing the relational database.

Throughout the early 1970s, application development focused on processing and reporting on frequently updated data and came to be known as online transaction processing (OLTP). Software development was predicated on a need to capture and report on specific KPIs that the business or organization needed. Though transistors and integrated circuits greatly increased the capabilities of these systems and started to bring down the cost of computing, mainframes and software were still too expensive to do much experimentation.

All of that changed with the introduction of lower cost minicomputers and then personal computers during the late 1970s and early 1980s. Spreadsheets and relational databases enabled more flexible analysis of data in what initially were described as decision support systems. But as time went on and data became more distributed, there was a growing realization that inconsistent approaches to data gathering led to questionable analysis results and business conclusions. The time was right to define new approaches to information architecture.

# The Data Warehouse

Bill Inmon is often described as the person who provided the first early definition of the role of these new data stores as "data warehouses". He described the data warehouse as "a subject oriented, integrated, non-volatile, and time variant collection of data in support of management's decisions". In the early 1990s, he further refined the concept of an enterprise data warehouse (EDW). The EDW was proposed as the single repository of all historic data for a company. It was described as containing a data model in third normal form where all of the attributes are atomic and contain unique values, similar to the schema in OLTP databases.

Figure 1-2 illustrates a very small portion of an imaginary third normal form model for an airline ticketing data warehouse. As shown, it could be used to analyze individual airline passenger transactions, airliner seats that are ticketed, flight segments, ticket fares sold, and promotions / frequent flyer awards.

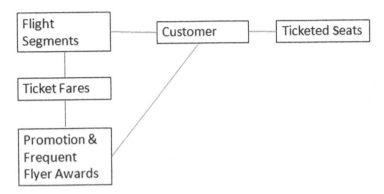

***Figure 1-2.*** *Simple third normal form (3NF) schema*

The EDW is loaded with data extracted from OLTP tables in the source systems. Transformations are used to gain consistency in data definitions when extracting data from a variety of sources and for implementation of data quality rules and standards. When data warehouses were first developed, the extraction, transformation, and load (ETL) processing between sources and targets was often performed on a weekly or monthly basis in batch mode. However, business demands for near real-time data analysis continued to push toward more frequent loading of the data warehouse. Today, data loading is often a continuous trickle feed, and any time delay in loading is usually due to the complexity of transformations the data must go through. Many organizations have discovered that the only way to reduce latency caused by data transformations is to place more stringent rules on how data is populated initially in the OLTP systems, thus ensuring quality and consistency at the sources and lessoning the need for transformations.

Many early practitioners initially focused on gathering all of the data they could in the data warehouse, figuring that business analysts would determine what to do with it later. This "build it and they will come" approach often led to stalled projects when business analysts couldn't easily manipulate the data that was needed to answer their business questions. Many business analysts simply downloaded data out of the EDW and into spreadsheets by using a variety of extractions they created themselves. They sometimes augmented that data with data from other sources that they had access to. Arguments ensued as to where the single version of the truth existed. This led to many early EDWs being declared as failures, so their designs came under reevaluation.

■ **Note** If the EDW "build and they will come" approach sounds similar to approaches being attempted in IT-led Hadoop and NoSQL database projects today, the authors believe this is not a coincidence. As any architect knows, form should follow function. The reverse notion, on the other hand, is not the proper way to design solutions. Unfortunately, we are seeing history repeating itself in many of these Big Data projects, and the consequences could be similarly dismal until the lessons of the past are relearned.

As debates were taking place about the usefulness of the EDW within lines of business at many companies and organizations, Ralph Kimball introduced an approach that appeared to enable business analysts to perform ad hoc queries in a more intuitive way. His star schema design featured a large fact table surrounded by dimension tables (sometimes called look-up tables) and containing hierarchies. This schema was popularly deployed in data marts, often defined as line of business subject-oriented data warehouses.

To illustrate its usefulness, we have a very simple airline data mart illustrated in Figure 1-3. We wish to determine the customers who took flights from the United States to Mexico in July 2014. As illustrated in this star schema, customer transactions are in held in the fact table. The originating and destination dimension tables contain geographic drill-down information (continent, country, state or province, city, and airport identifier). The time dimension enables drill down to specific time periods (year, month, week, day, hour of day).

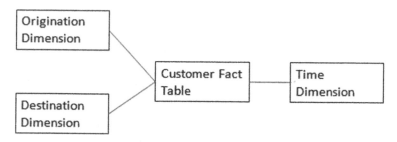

*Figure 1-3.* *Simple star schema*

Not all relational databases were initially adept at providing optimal query performance where a star schema was defined. These performance challenges led to the creation of multidimensional online analytics processing (MOLAP) engines especially designed to handle the hierarchies and star schema. MOLAP engines performed so well because these "cubes" consisted of pre-joined drill paths through the data. Figure 1-4 pictures a physical representation of a three-dimensional cube.

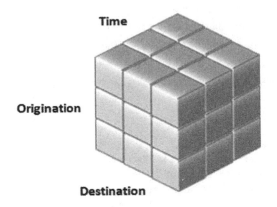

*Figure 1-4.* *Three-dimensional cube representation*

Later, as relational database optimizers matured, it became possible to achieve good query performance when deploying the star schema within the relational database management system. These became known as relational online analytical processing (ROLAP) implementations.

## Independent vs. Dependent Data Marts

In the mid-1990s, there was much debate about the usefulness of the EDW when compared to data marts. When business analysts found the star schema was easier to navigate (and often deployed their own marts), some IT database programmers responded by creating views over the top of the data and schema in their EDW to overcome this objection. However, the programming and maintenance effort in building views was typically not timely enough to meet the growing demands of business analysts.

Another problem often arose. When individual data marts are defined and deployed independently of each other and don't follow data definition rules established within the EDW, inconsistent representation of the common data can call into question where the true data is housed. Figure 1-5 illustrates the complexity that can emerge when various lines of business build their own independent data marts and extract data directly from OLTP sources. In actual implementations, the complexity is sometimes greater than what is shown here as data might flow directly between data marts as well. Spreadsheets might also be added to this illustration serving as business intelligence tools tied to unique storage and representations of data. Organizations that deploy in this manner generally spend a great amount of time in business meetings arguing about who has the correct report representing the true state of the business, even if the reports are supposed to show the same KPIs.

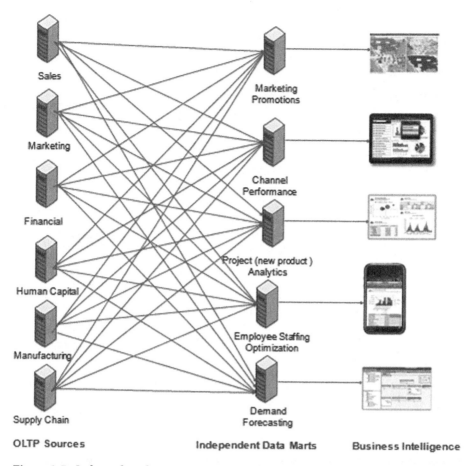

***Figure 1-5.*** *Independent data marts with unique ETL between sources and target marts*

In the end, the debate should not have been about EDWs vs. data marts. There were solid reasons why both approaches had merit where the right architectural guidelines were applied. As many information and data warehouse architects began to realize this, a blended approach became the best practice. EDWs were implemented and extended incrementally as new sources of data were also required in the data marts. The data marts were made dependent upon the data definitions in the EDW. As the EDW remains the historic database of record, data fed into the marts is extracted from the EDW. The exception to using the EDW as the source of all data typically occurred when there was unique third-party data that was relevant to only a single line of business in an organization. Then that unique data was stored only in that line of business's data mart.

Figure 1-6 illustrates data marts dependent on the EDW. This approach often leads to defining conformed dimensions to establish consistency across data marts. When conformed dimensions are defined, it is possible to submit a single query that accesses data from multiple data marts (since the dimensions represent the same hierarchies in the various data marts).

7

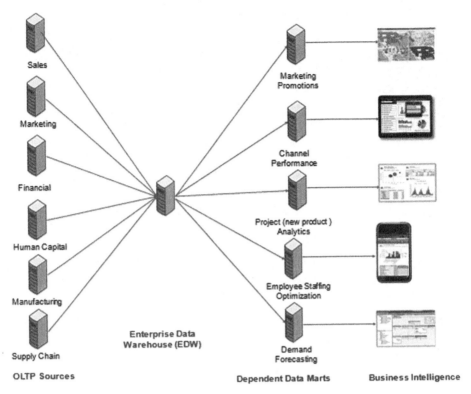

**Figure 1-6.** *Dependent data marts with ETL from the EDW, the trusted source of data*

Database data management platforms you are most likely to encounter as data warehouses and / or data mart engines include the following: Oracle (Database Enterprise Edition and Essbase), IBM (DB2 and Netezza), Microsoft SQL Server, Teradata, SAP HANA, and HP Vertica. ETL tools that are commonly deployed include Informatica, Oracle Data Integrator, IBM DataStage, and Ab Initio.

---

■ **Note**  When the EDW and data marts first became central and mandatory to running the business, information architects began to understand the need for these platforms to also be highly available, recoverable, and secure. As Hadoop clusters and NoSQL databases are assuming similar levels of importance to lines of business today, the demand for similar capabilities in these platforms is driving the creation of new features and capabilities in these distributions. This is illustrated by the growing focus on improved availability, recoverability, and security in the more recent software releases in the open source community and being offered by the various vendors creating distributions.

---

# An Incremental Approach

Early data warehousing design efforts sometimes suffered from "paralysis by over analysis" with a focus on elegant IT designs but not mapped to requirements from lines of business. Designs of early EDWs often took 12 months or more, well outside the bounds of business needs or the attention spans of business sponsors. Some early practitioners relied on a classic waterfall approach where the scope of the effort for the entire EDW was first determined, and then time and resources were allocated.

Figure 1-7 illustrates the waterfall approach. Lengthy project plans, delays, and lack of attention to the business often led to the lines of business taking matters into their own hands, developing and deploying independent data marts, or creating pseudo data marts in spreadsheets to solve their most immediate problems.

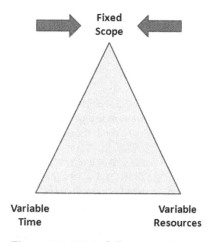

*Figure 1-7. Waterfall approach*

In light of these problems, many turned away from the waterfall approach and switched to an agile incremental approach to design and development. Partnerships were formed between IT and the lines of business. Time frames of 120 days or less for implementation and evaluation of the progress toward a business solution became commonplace in many organizations. Figure 1-8 represents the incremental approach and illustrates a fixed time and fixed resources being assigned.

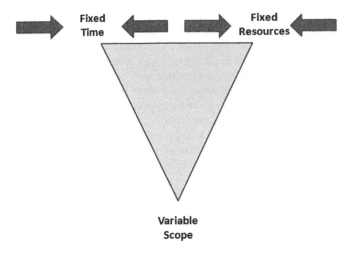

*Figure 1-8. Incremental approach*

Though Figure 1-8 shows a variable scope, there should be some real business value demonstrated that is aligned to the business goals at each increment in the process. So, in practice, the methodology applied is often a blended balancing of the incremental and waterfall approaches.

Using this approach, the usefulness of the solution is re-evaluated at each step along the way. Defining and evaluating the EDW and dependent data marts in shorter time increments means that IT designs and development can be adjusted before they became too misaligned with business expectations. Return on investment can be calculated at regular intervals and reflect any changes in scope.

Some companies choose to embed business analysts within their IT organizations to drive ongoing identification of incremental requirements. Others create business intelligence centers of excellence as virtual organizations, periodically bringing together analysts in the lines of business with IT architects and developers. Ongoing communications and flexibility among the teams is critical to success regardless of the approach used.

---

■ **Note**  Successful Big Data projects that provide new business solutions are also usually developed using an incremental approach. Ongoing dialog in the organization among the teams regarding what is being discovered and the potential business impact is essential to success.

---

## Faster Implementation Strategies

Early data warehouse implementations were based on entirely customized designs. Since data models had not matured and were not widely available, a significant amount of time was spent defining and designing the data models from scratch. Workload characteristics and workload changes over time were very unpredictable, making the specification of servers and storage difficult. As lessons were learned and data warehouse designs matured, best practices became understood and solutions emerged that built upon these experiences.

One set of solutions to emerge were predefined data models based on common data analysis needs. The models became available for industries (such as retail, communications, and banking) and provided horizontal analytic solutions (for example, financials, sales, marketing, and supply chain). Such models are available today from software vendors and consulting companies and feature definitions of logical designs and sometimes also include physical schema. They cover key business areas and contain the tables and other data elements needed to start the deployment of the data warehouse. Some are also packaged with ETL scripts useful in extracting data from popular ERP and CRM transaction processing sources and loading the data into the data models. Of course, most organizations customize the models based on their own unique business requirements. However, the models do provide the starting point for many data warehousing projects and are most successfully deployed when using the incremental approach.

As we noted earlier, configuring servers and storage for data warehousing workloads also presented challenges for information architects and server and storage architects. Given that data volumes grow at a much more rapid rate than the evolution of faster access times to physical disk drives, most found their platforms became throughput-bound if not enough attention is paid to the overall system design. In recent years, the notion of deploying appliance-like platforms configured for data warehousing and data marts has become quite common. There are several such offerings available from relational database vendors who also provide servers and storage. The availability of flash in storage further helped speed performance where the database software was optimized to take advantage of the flash. More recently, the dramatic reduction in the cost of memory, introduction of new processors capable of addressing huge memory footprints, and further refinement in the database's ability to store and retrieve frequently accessed data in-memory led to huge query response and analysis performance improvements. All of these have served to mitigate many of the complex database tuning and design tasks previously necessary.

That said, as certain server and storage bottlenecks such as throughput are overcome, others will naturally arise since there is always a physical limitation somewhere in a system. Business analysts will continue to demand new analytic applications that take advantage of new platform capabilities and push the bounds of the technology.

---

■ **Note**  At the time of publication, the number of applications for Hadoop clusters and NoSQL databases was quite small but growing. There were also a growing number of appliance-like server and storage platforms available for these data management engines. As the business value of the solutions that require such engines is understood, time to implementation and the ability to meet service levels will grow in importance. So, it is expected that the desire for such optimally configured appliances will grow and their popularity will follow a trajectory similar to what was observed in the adoption of data warehousing appliances.

---

# Matching Business Intelligence Tools to Analysts

How data is accessed and utilized is driven by the needs and skills of the individuals in the lines of business. For those who need to see the data to make decisions, the tools they might use can range from simple reporting tools to extremely sophisticated data mining tools. Modern infrastructures sometimes also include engines for automated recommendations and actions, as well as information discovery tools.

Figure 1-9 illustrates the range of tools and techniques and their relative user community sizes and relative complexity.

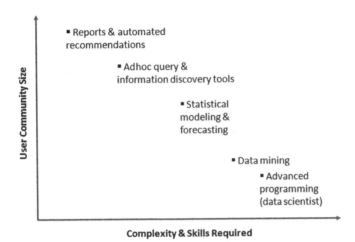

**Figure 1-9.** *Potential business analyst and user community size for various analyst tools*

The simplest way to convey information to business analysts is through pre-defined reports that display KPIs selected by developers of the reports. Reports have limited flexibility in the information that can be viewed, but they also assure that a wide variety of business users can become consumers of the information because of the simplicity in accessing them. The reporting tools the developers use generate SQL for accessing needed data. Report developers often judge the quality of reporting tools by the clarity with which they present the KPIs and the ease and flexibility in which reports can be generated, shared, and distributed. For example, a variety of template types are often supported for printing such as PDF, RTF, and XSL.

Ad hoc query and analysis tools provide a greater degree of flexibility since business analysts can pose their own *what-if* questions by navigating database tables themselves. Developers create business metadata to translate cryptic table names into meaningful business-oriented data descriptions. The ease with which business users can navigate the data is also dependent on the underlying schema design in the database. As we described earlier, star schema with dimensional models and hierarchies are particularly easy to navigate. Figure 1-10 illustrates an interface showing a fact table consisting of sales surrounded by dimensions that include time, products, offices, and others. Behind the interface, these tools also generate SQL to access the data. In addition to flexibility, modern ad hoc query and analysis tools are judged by the data visualization capabilities these tools provide.

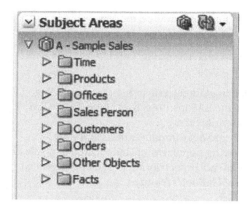

***Figure 1-10.*** *Typical ad hoc query tool interface to facts and dimensions*

Typical ad hoc query, analysis, and reporting tools you are likely to see being used today include Oracle Business Intelligence Foundation Suite, SAP Business Objects, IBM Cognos, MicroStrategy, Tableau, QlikView, Pentaho, and Tibco Jaspersoft. Of course, many would say that Microsoft Excel is the most popular tool for this type of work in their organization. In most organizations, a variety of vendors' tools are used.

A smaller but growing subset of business analysts deal with massive amounts of data and seek to uncover hidden patterns and / or predict future outcomes using their data. The kinds of analyses range from the simple statistics that you probably learned in college (for example, mean, standard deviation, and so on) to models based on more sophisticated data mining algorithms.

The statistical functions that business analysts work with to bring clarity to the data usually fit in the following categories:

- Basic statistical functions such as summary, sort, rank, and frequency

- Density, probability, and quantile functions

- Special functions such as gamma functions

- Test functions such as chi square, simple and weighted kappas, and correlation

Advanced data mining algorithms are used when there is a need to understand what variables are critical in accurately predicting outcomes and in defining the predictive models that will subsequently be used to predict the outcomes. The models are often applied where there are hundreds of variables present but only a dozen or fewer that impact the outcome. The data mining algorithms can be categorized as follows:

- Clustering algorithms: Used to explore where certain business outcomes fall into to certain groups with common characteristics such as teenagers, males, and so on.

- Logic models: Used where if certain events occur, others will follow (and often referenced as decision trees).

- Neural networks: Somewhat black box mathematical models trained against sample sets with known outcomes.

- Anomaly detection algorithms: Used to detect outliers and rare events.

The vendors you are likely to find installed in your current architecture providing statistical and data mining capabilities include the SAS Institute, IBM SPSS, R (an open source statistical engine), and Oracle Advanced Analytics.

Historically, statisticians and data miners were also domain experts and were sometimes referred to as "quants." With the growing popularity of Hadoop, the new role of data scientist has emerged. Early data scientists were especially adept at using advanced programming techniques that took advantage of Hadoop's features.

---

▪ **Note**   There is much debate today about the skills and definition of the data scientist role. Some still believe the data scientist is a combination of a statistician and Hadoop programming guru. However, many hired with those skills have shown that they lack the domain expertise needed to understand what to look for in the data and the potential impact on the business. In many organizations today, data scientists are paired with business domain experts, and they work as a team to assure success.

---

Early in this century, it was recognized that there was a growing need to explore massive data sets that might include structured, semi-structured, and streaming data. The information discovery tools that were introduced enable exploration of data where a schema is not pre-defined. The tools generally either have their own proprietary data store engines, such as Oracle Endeca Information Discovery, or rely on Hadoop to enable exploration of data sets and combinations of data. The data analyzed is typically gathered from OLTP sources, EDWs, NoSQL databases, and Hadoop. Tibco Spotfire, Oracle Big Data Discovery, and some of the business intelligence tools we previously mentioned in this chapter can directly access Hadoop and are used for information discovery.

Finally, for certain problems, action must be taken in real time. Examples might include recommending products that could be of interest during a web site shopping visit or equipment that should be checked out for maintenance because its failure is predicted in the near future.

Web site activity data is typically analyzed using predictive analytics models. The models' results are periodically provided as updates (using batch feeds) to a real-time recommendation engine. The engine then recommends that the web site serve up specific web pages or notifications as guided by the models. As more analyses are made, the recommendations are fine-tuned and become more accurate. Often, reporting tools are used to monitor the results of these automated actions.

Other business problems, such as pending equipment failure, might require immediate action prior to any detailed data analysis since there is latency in the previously described learning process. Business rules engines or event processing engines can be pre-programmed to take specific action as a result of detected events.

These are often deployed in Internet of Things solutions in order to trigger an immediate action based on what sensors are detecting.

Later in this book, we will describe how to uncover the need for these various tools and solutions and then subsequently describe technical considerations as they become part of the information architecture design.

# Evolving Data Management Strategies

As business workload demands changed and drove new technical requirements, relational databases evolved and introduced new capabilities intended to address those requirements. However, some found that a technology based on a concept of data's fitting neatly in rows and columns introduced too much overhead or was misaligned with the problems that needed to be solved. It is largely for those reasons that NoSQL databases and Hadoop engines began to appear around the turn of this century.

Coincidentally, they appeared at a time when the "open source" movement was gaining momentum and, in turn, helped to fuel that momentum. In the open source model, vendors and individuals have access to source code and these "committers" submit updates and utilities they are willing to share. Source code for NoSQL databases can be licensed from the Apache Software Foundation and GNU. Hadoop licenses can be obtained from Apache. As new features are incorporated into new releases of the open source code, the software vendors then determine what to include in their own distributions. Though the distributions can be downloaded for free, the vendors believe they can ultimately become profitable and successful companies by generating revenue through subscriptions (including support) and by offering services for a fee.

## NoSQL Databases

The NoSQL database terminology dates to the late 1990s and was intended to describe a broad class of non-relational database engines designed to handle rapid updates and ingest the largest quantities of data while providing horizontal scalability. Such update and ingestion workloads had become a challenge for certain online applications (such as shopping carts on web sites) where fast update performance was critical despite a huge number of users of the application.

Early NoSQL databases did not support SQL, hence the name for this class of data management engines. Over time, SQL support of varying degrees has been added to many of the available NoSQL databases. Early NoSQL databases also did not provide traditional atomicity, consistency, isolation, and durability (ACID) properties provided by a relational database. This support was deemed as undesirable since it required too much overhead that got in the way of the performance needed. Today, many of the NoSQL databases are claiming to support at least some of the ACID properties. However, it is generally recognized that they are not intended to be used as a substitute for OLTP relational database engines or where joining many types of data across dimensions is required.

A variety of NoSQL database types have emerged. These include the following:

- Key Value Pairs: Databases that consist of keys and a value or set of values and that are often used for very lightweight transactions and where the number of values tied to a key grows over time.

- Column-based: Databases that are collections of one or more key value pairs, sometimes described as two-dimensional arrays, and are used to represent records so that queries of the data can return entire records.

- Document-based: Similar to column-based NoSQL databases, these databases are designed for document storage and feature deep nesting capabilities, enabling complex structures to be built such that documents can be stored within documents.

- Graph-based: Databases that use treelike structures with nodes and edges connected via relations.

Horizontal scalability of NoSQL databases is enabled using a technique called *sharding*. Sharding is simply the spreading of data across multiple independent servers or nodes in a cluster. Performance is dependent upon the power of the nodes but also upon how well the spreading of the data provides a distribution that also matches the performance capabilities of the individual servers. For example, if all of the most recent data is put on a single node and most of the activity is related to recent data, the application will not scale well. Many NoSQL database vendors have focused on automating the sharding process to provide better load balancing and make it easier to add or remove capacity in the cluster.

Though not as robust as relational databases in supporting high availability failover scenarios, NoSQL databases do enable replication of data to provide database availability in case of server or node failure. Copies of data are typically replicated across nodes that are different from the nodes where the primary data resides.

There are dozens of NoSQL database engines of the various types we have described. Some that you are more likely to encounter include Apache Cassandra, MongoDB, Amazon DynamoDB, Oracle NoSQL Database, IBM Cloudant, Couchbase, and MarkLogic. As the feature list for these databases can rapidly change, understanding the capabilities that are provided by the version being considered for deployment is very important. As an example, some added in-memory capabilities in only their more recent versions.

# Hadoop's Evolution

Streaming data feeds from web sites increasingly caused headaches for companies and organizations seeking to analyze the effectiveness of their search engines at the beginning of this century. Such data streams include embedded identifiers and data of value, but also other miscellaneous characters. Figure 1-11 provides an illustration of typical data found in web logs. Clearly, this type of data does not fit well in a relational database. Doug Cutting was working on an approach to solve this problem by developing a new engine that he called *Nutch* as early as 2002. In 2003 and 2004, Google published two important papers describing the Google File System (GFS) and MapReduce. The notion of a distributed file system was not new at the time, but Google's papers laid out a vision of how to solve the search problem.

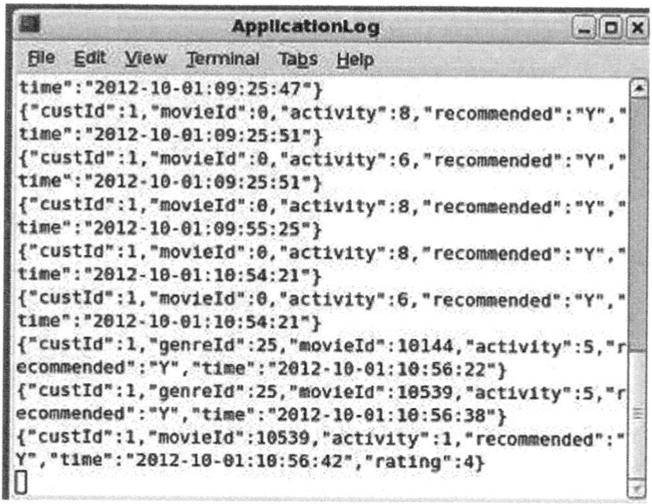

```
ApplicationLog                                  _ □ x
File  Edit  View  Terminal  Tabs  Help
time":"2012-10-01:09:25:47"}
{"custId":1,"movieId":0,"activity":8,"recommended":"Y","
time":"2012-10-01:09:25:51"}
{"custId":1,"movieId":0,"activity":6,"recommended":"Y","
time":"2012-10-01:09:25:51"}
{"custId":1,"movieId":0,"activity":8,"recommended":"Y","
time":"2012-10-01:09:55:25"}
{"custId":1,"movieId":0,"activity":8,"recommended":"Y","
time":"2012-10-01:10:54:21"}
{"custId":1,"movieId":0,"activity":6,"recommended":"Y","
time":"2012-10-01:10:54:21"}
{"custId":1,"genreId":25,"movieId":10144,"activity":5,"r
ecommended":"Y","time":"2012-10-01:10:56:22"}
{"custId":1,"genreId":25,"movieId":10539,"activity":5,"r
ecommended":"Y","time":"2012-10-01:10:56:38"}
{"custId":1,"movieId":10539,"activity":1,"recommended":"
Y","time":"2012-10-01:10:56:42","rating":4}
```

*Figure 1-11.* *Typical web log data stream*

Cutting understood the importance of the Google papers and made modifications to his own effort. MapReduce was able to map the data streams and reduce the data in the streams to data of value. GFS provided clues on how to scale the engine and such scalability was seen as particularly critical given the number of deployed web sites was exploding. In 2006, Cutting joined Yahoo! and renamed his storage and processing effort after the name of his son's toy elephant. Hadoop was born. That same year, Hadoop became an Apache Software Foundation project.

A distributed file system enables highly parallel workloads to occur across massive amounts of storage. MapReduce is co-located with the data providing the scalability needed. When this combination was discussed by early proponents, it was often described as solving a Big Data problem where data had huge volume, variety, and velocity. Over time, the Big Data terminology has taken on much broader meaning as vendors have positioned many different solutions to address many different though similar problems.

Today, Hadoop clusters are seen as the ideal solution for processing many types of workloads. Some of these clusters are now used to speed ETL processing by providing highly parallelized transformations between source systems and data warehouses. Other Hadoop clusters are used for predictive analytics workloads as analysts use tools such as R or SAS or leverage Hadoop's own machine learning and data mining programming library named Mahout. Data scientists sometimes write code (using Java, Python, Ruby on Rails, or other languages) and embed MapReduce or Mahout in that code to uncover patterns in the data. Increasingly, many also access data in the cluster through various SQL interfaces such as Hive, Impala, or other similar vendor offerings.

> ■ **Note**  Newer data management solutions described in this book were invented to provide optimized solutions by addressing specific emerging workload needs. However, the perception about many of these data stores, promoted as open source, is that they are cheaper. This can lead to the application of these software distributions outside of their technical sweet spots in order to reduce cost of acquisition and support. As developers complain about the limited functionality compared to other engines that were seen as more costly, vendors that create the software distributions often add new features and capabilities in response. The unanswered question is whether many of the resource utilization and performance benefits of these distributions will disappear as they overlap more with other legacy data management solutions and with each other.

# Hadoop Features and Tools

The Apache Software Foundation provides incubators for Hadoop features and tools and classifies these as development projects. As new releases occur, the results of these projects make their way, in varying degrees, into Hadoop distributions from vendors that include Cloudera, Hortonworks, IBM, MapR, and Pivotal. Apache Hadoop project status updates are posted on the apache.org web site.

If you are new to Hadoop, some definitions of key projects that the distributors and developers often talk about as Hadoop technologies and features could be useful. Some of the core data management features include the following:

- HDFS: The Hadoop Distributed File System.

- Parquet: A compressed columnar storage format for Hadoop.

- Sentry: A system that enables fine-grained, role-based authorization to data and metadata stored in Hadoop.

- Spark: An engine that enables Hadoop in-memory data processing.

- YARN: A framework used in scheduling and managing jobs and cluster resources.

- Zookeeper: A coordination service for distributed applications.

Important features for data transfer and collection in Hadoop include the following:

- Flume: A service for collecting and aggregating streaming data including log and event data in HDFS.

- Kafka: A publish-and-subscribe message broker platform designed to handle real-time data feeds.

- Sqoop: A tool used to transfer data between Hadoop and databases.

Programming tools, application programming interfaces (APIs), and utilities in Hadoop include the following:

- Hive: A SQL-like construct (HiveQL) for querying data in Hadoop.

- MapReduce: An early Hadoop programming paradigm that performs a "map" (filtering and sorting) and then a "reduce" (summary operation) for data that is distributed across nodes.

- Oozie: A workflow job scheduler used in managing Hadoop jobs.

- Pig: A data-flow language and parallel execution framework for data processing.

- Spark GraphX: An API that enables viewing of data as graphs and collections, transformations, and joins of graphs to resilient distributed data sets (RDDs), and creation of custom graph algorithms in Spark.

- Spark MLib: A machine learning library implemented in Spark.

- Spark SQL: An API that enables querying of RDDs in Spark in a structured (Hive) context.

- Spark Streaming: An API that enables creation of streaming processes in Spark.

- Solr: A full text indexing and search platform.

Creators of Hadoop-based applications sometimes seek the capabilities provided by a NoSQL database as part of their designs. HBase provides a NoSQL columnar database that is deployed on HDFS and enables random reads and writes. It is especially useful in handling sparsity of data problems. In addition to supporting ad hoc queries, HBase often is used for providing data summaries.

Layout of a Hadoop cluster on the underlying servers and storage requires the designation of name nodes, data nodes, and nodes that will provide the services enabling the features that we previously mentioned. Proper deployment of services across the cluster eliminates critical single points of failure that could bring the entire Hadoop cluster down. Data is normally triple replicated to assure that it is available in the event of node failures.

Some debate remains about when it is appropriate to include a Hadoop cluster as a component in the information architecture as opposed to suggesting a data warehouse deployed using a relational database. Table 1-1 attempts to highlight the strengths of each. As the capabilities in the data management engines are rapidly changing, you should revalidate these characteristics based on the most current information available when you consider deployment options for projects of your own.

*Table 1-1.* *Summary of Some Differences When Deploying Hadoop vs. Data Warehouse (Relational Database)*

| Characteristic | Hadoop | Data Warehouse |
|---|---|---|
| Data Value | Data usually of mixed quality & value—volume most important | Data only of high quality value most important |
| Schema | Most often used as distributed file system | Most often 3NF & star schema hybrid |
| Typical Workloads | Information discovery, predictive analytics, ETL processing | Historic transactional reporting, ad-hoc queries, OLAP |
| Data Sources | Varied data types from streaming to structured | Structured data sources |
| Availability | Data replication across nodes | Guaranteed consistent failover |
| Security | Authentication, encryption, access control lists | Same as Hadoop plus even finer granular control |
| Scalability | Can be distributed over 100s nodes, 100s petabytes data | Multi-node, typically 100s terabytes or a few petabytes of data |

# The Internet of Things

Within this decade, the growing popularity of reporting on and analyzing data gathered from sensors and control devices became clear. Speculation about the value of such solutions and early testing of this idea began as early as the 1980s with experimental vending machines connected to the Internet. Early this century, there was much speculation about machine-to-machine (M2M) communications and the value such capabilities could provide. There were even jokes about how your kitchen appliances might be plotting against you. However, it soon became clear that the solution footprint would involve more than just the devices themselves, and so the *Internet of Things* terminology was added to our vocabulary.

Prior to the invention of this catch phrase, many manufacturers of devices had already added the capability to gather data from their devices by outfitting them with intelligent sensors and controllers. Some included early versions of simple data gathering software. However, the price, size, and limited functionality of early sensors and controllers sometimes also limited the usefulness of the data that could be gathered. Further, the potential business usage of such data was not widely understood, and the software needed to analyze it was not optimal. So, when the data was gathered at the source, often it was not transmitted or analyzed. The data was simply thrown away.

Since design of manufactured products requires lengthy lead times and the products can have significant lifetimes once they are produced, engineers continued to add sensors and the capability to gather intelligence to the products they were designing. They foresaw a day when the data could be useful in multiple ways. Many understood that such data could be utilized to better understand product quality and the potential failure of components, enable automated requests for service, provide information on environmental factors, aid in better energy management, and provide data for hundreds of other potential applications.

---

■ **Note** Technology zealots sometimes ask us to describe the "killer user cases" for Big Data (and specifically Hadoop) in the industry that they work in. Where sensors and intelligent controllers increasingly provide data that is critical to running the business, the use case they are seeking can become readily apparent. Their Big Data use case could be driven by the need to analyze data from the Internet of Things.

---

The growing demand for sensors led to more research that led to further sensor miniaturization and significant reductions in price. Miniaturization was partly enabled by a huge reduction in energy needed to power the sensors. As a result, billions of sensors are deployed today, and this number is expected to soon grow to hundreds of billions deployed by the end of the decade. That growing volume at a lower cost will continue to drive further innovation and momentum for their use.

Fortunately, a second breakthrough was occurring as sensors and intelligent controllers proliferated. This breakthrough was in the capability of newer types of data management software to ingest and analyze huge volumes of streaming data. Though NoSQL databases and Hadoop initially were most often deployed to process and analyze web site traffic and social media data, it turned out that these engines are also ideal for the capture and analysis of data streams coming from sensors and controllers.

Today, they are used to gather and analyze streaming data from automobiles, jet engines, mobile devices, health monitors, items in shipment, and many other sources. Figure 1-12 illustrates a simplified view of key Internet of Things components. Our information architecture introduction earlier in this chapter focused on the data management and business intelligence platforms on the right side of the illustration. The Internet of Things further introduces a need for data routing and event processing, provisioning and management of the software on the sensors, identity access controls for securing data transmissions in the middle tier, and an appropriate communications network from the sensors and devices to the corporate intranet infrastructure. Still challenging to such implementations, at time of publication, were the emerging and competing standards and consortia addressing these areas. Among those weighing in were the Open Internet Consortium (end-to-end Internet of Things architecture), IETF (for communications and encodings), the AllSeen Alliance (for proximal device connectivity), the IPSO Alliance (for data representation), the Open Mobile Alliance (for device management and object registries), and the Thread Group (for smart home networks). The Industrial Internet Consortium was seeking to standardize vertical solutions for industrial applications and a variety of industry groups were also attempting to define standards.

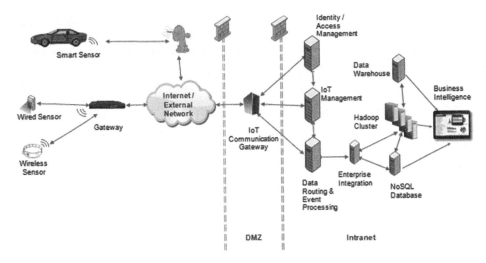

**Figure 1-12.** *Simplified view of Internet of Things components*

Various strategies have emerged in organizations that want to analyze data from the Internet of Things. Some design and develop the entire pictured footprint. Others design and deploy the smart sensor code (usually programming it in Java), manage it, and then focus on gathering and analyzing the data while partnering with communications infrastructure providers to enable and secure transmissions. Still others only focus only on analyzing the data, relying on others to provide communications and build out intelligence in the sensors.

Our description of extending the information architecture to include NoSQL databases and Hadoop clusters will be relevant to all three of these scenarios. But we will also later describe in this book some of the additional components unique to the Internet of Things and additional decisions you might have when considering communications, security, and provisioning of intelligent sensors.

# The Methodology in This Book

The remainder of this book describes a methodology for developing and deploying projects. The authors observed this methodology as commonly used when previous generation projects succeeded. It is now being applied and extended to Big Data and Internet of Things efforts.

The methodology we describe is not revolutionary. It is built upon accepted best practices that most enterprise architects are familiar with. We realize that the audience for this book is much wider than those with an architecture background, so we'll next describe the popular The Open Group Architectural Framework (TOGAF) model from The Open Group and how we can apply it to our methodology. Especially important when applying any methodology in the creation of the next generation information architecture is to use the incremental approach we earlier described in this chapter.

# TOGAF and Architectural Principles

The Open Group is a worldwide organization offering standards and certification programs for enterprise architecture. First established in 1995, TOGAF has been widely adopted and provides the basis for architectural design methodologies present in many of its member organizations. At the time of publication, there were over 350 organizations taking part as members. Many take part in The Open Group boards, councils, member forums, work groups, or technical work groups that define best practices and standards.

TOGAF itself describes four types of architecture. All are applicable to defining the information architecture we describe in this book. The four types are as follows:

- Business architecture comprised of business strategy, governance, organization, and key business processes

- Data architecture consisting of logical and physical data asset structures and data management resources

- Application architecture describing how applications are deployed, how they interact with each other, and how they relate to business processes defined by the business architecture

- Technology architecture describing the logical software, server, storage, networking, and communications capabilities required

Business architecture is often the most overlooked of the four when IT specialists and architects define and develop the information architecture, but it is extremely important to achieving overall success in these projects. We spend some time in this book describing the uncovering of a company's or organization's business strategy and their processes that are critical to running the business. The organization of the business clearly can drive various aspects of the technical information architecture, such as where data marts might be required to meet the needs of specific lines of business. The topic of data governance is top of mind in both business and IT, and the topic appears in many chapters.

At the heart of information architecture is the data architecture and technology architecture. You will see us focus on these areas in chapters that discuss gathering IT requirements and the future state design in particular. The application architecture described by TOGAF defines the business solutions in our information architecture and the relationship these solutions have to each other when built upon the underlying data and technology architecture components.

You or your architecture team might be most familiar with TOGAF in the context of the standard that it provides. The TOGAF standard consists of the following parts:

- An introduction to key concepts in enterprise architecture and TOGAF

- An architecture development method (ADM) that describes a step-by-step approach to developing the enterprise architecture

- A collection of ADM guidelines and techniques for applying it

- An architecture content framework including reusable architecture building blocks (ABBs) and typical architecture deliverables

- An enterprise continuum that provides a model for classifying artifacts and showing how they can be reused and modified over time

- TOGAF reference models including a technical reference model (TRM) and integrated information infrastructure model (III-RM)

- An architecture capability framework providing guidelines, templates, and resources useful when establishing the architecture practice in an organization or company

It is no coincidence that the scope of TOGAF maps extremely well to the information architecture methodology we describe in this book since the methodology is based upon lessons learned using standard architectural techniques derived from TOGAF. If you seek more details on TOGAF than we cover here, we strongly recommend you explore The Open Group's web site at www.opengroup.org/TOGAF. Becoming a member can introduce you to a wealth of information and, if you are an architect, provide you with opportunities to become a contributing member of a community providing architecture thought leadership.

---

■ **Note**  The IT audience for this book might also wonder about the applicability of ITIL (formerly known as the Information Technology Infrastructure Library) when it comes to defining an information architecture. ITIL is closely linked to ISO/IEC 20000, an international standard for IT service management, and defines a framework and certification process. More recently, ongoing development of ITIL came under the direction of AXELOS (www.axelos.com), a company co-created by the UK Cabinet Office and Capita PLC. The five major service areas that ITIL addresses are service strategy, service design, service transition, service operation, and continual service improvement. The level of IT services provided must align with the business needs of your organization and these in turn should guide you when defining the future state technical architecture that is described later in this book. So, ITIL is complementary to TOGAF as it can be used to help you define how you will operationalize the architected future state.

---

## Our Methodology for Success

The methodology we focus on in the subsequent chapters of this book consists of seven phases ranging from an early conceptual vision through project implementation. Each phase is represented by its own chapter in this book.

The first phase we describe in detail establishes an early vision of the future state information architecture. In the next phase, we determine the business drivers and key performance indicators required. Then we map the KPIs and key measures to sources for the data and determine how the data will be provided as usable information. We next assess skills we have available in our organization. Based on the information we gathered in the previous phases, we can then design in much more detail the future state information architecture. We next define and agree upon a roadmap describing the implementation sequence of our future state architecture. Then we are ready to begin implementation.

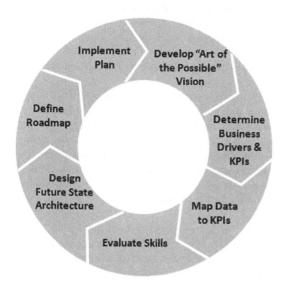

**Figure 1-13.** *Phases in our methodology for success*

Figure 1-13 illustrates this process. It is shown as a continuous circle to indicate that it does not end with implementation. Let's take a quick look at what happens in each of these phases.

---

■ **Note**   In practice, the phases of the methodology sometimes overlap. For example, you might discover while working on a phase that critical information you need is missing and that the information should have been gathered in a previous phase. Of course, when that occurs, you must go back and gather the missing information in order to proceed with the current phase.

---

When we paint a vision of a future architecture, we start with a basic understanding of our current state and we begin to speculate on how it might evolve. A challenge in many organizations is that at inception, the technical vision may not be aligned to the business vision. In fact, business visionaries and potential sponsors must drive the technical vision. So we must understand current business utilization of our current information architecture and how that could change in the future. As we will describe in the next chapter, the vision phase is mostly about gathering requirements and exploring the art of the possible.

The next phase of the methodology takes a much deeper look at the business drivers. Line of business sponsors and business analysts provide more insight into what is required to run the business today and also address the new challenges they are facing. During this phase, we must gain an understanding of their critical success factors, the key performance indicators that the business analysts need, and the measures that must be gathered. All of these will drive the technical architecture design that follows. We also begin to understand the potential business benefits that will accrue by having access to this data and begin to prioritize different phases of the project based on their business value and opportunity for success.

Once we understand what data our business analysts need, we must figure out where we should get the data. Line of business sponsors and business analysts will help us here as well. They can describe the quality of the data needed, the sources that they trust, and data governance requirements. Furthermore, they can help us understand how the data will be analyzed, the granularity needed, how long it must be retained, and what form it should be delivered in (for example, as reports or in ad hoc query tools or through data exploration and visualization tools). They can also describe the infrastructure availability requirements that should be driven by the need for timely decisions to be made using the data and the potential impact on the business.

At this point, we should now understand the data and analysis requirements. We are ready to design our future state technical architecture and the IT architecture team will engage extensively in this phase. However, before a more detailed design is started, we might want to first understand the skills we have in our organization and the impact those skills (or lack of) might have in the architecture. We'll also want to clearly understand the good and bad things about our current state architecture and how we might extend it through the introduction of new software components and systems. Some initial notion as to the scale of costs needed to redefine our footprint should become apparent at this time.

A bill of materials for new hardware and software is useful, but to truly understand when we'll reach a positive return on investment, we must also begin to understand the potential implementation costs for the various phases envisioned. Implementation costs generally dwarf software and other infrastructure considerations. Based on skills gaps identified, we should begin to assess the cost of services from systems integrators to fill those gaps. Other factors such as scope of effort and risk of implementation are also evaluated.

Once we have this information, we can develop a high-level roadmap to implementation backed by a reasonable and understandable business case. Our target audience for the roadmap includes executives and sponsors who will fund the project. If we've done our job right throughout the process, there should be few surprises at this phase of the effort. Part of the presentation could be a mock-up demonstration of the business capabilities that will be delivered. Much of the dialogue should be about priorities and whether the project phases are in the right order. Executives and sponsors might also ask for more details regarding costs, and likely those will be directed at the cost of implementation. But with a solid business case, a go-ahead to proceed with the project is likely.

Lastly, there is the implementation itself. As noted previously, an incremental approach will assure the project appears to be tracking well and assumptions were correct. Along the way, subsequent phases may be reordered and / or pulled forward into initial phases based on changing business priorities and challenges uncovered during the implementation. All of this must be accomplished without falling victim to scope creep. In addition to demonstrating progress in solution delivery, tracking of costs of implementation and reporting on those at regular intervals also demonstrates accountability.

As the project reaches its initial completion and you deliver on the agreed upon blueprint, it is important to claim success. However, it is likely that the lines of business and IT will have already started to develop a revised vision of what comes next. These projects always evolve as the business needs change and as business analysts become more advanced in their understanding of what is possible. And so the cycle repeats itself again and again.

For the remainder of this book, we will take a much deeper look at each of these topics. We will begin by understanding the art of the possible as we define a vision of the future state information architecture.

# CHAPTER 2

■ ■ ■

# Evaluating the Art of the Possible

Fear of being left behind can be powerful motivation. Today, many organizations embark on building Big Data and Internet of Things prototypes simply because they fear that their competitors are gaining a head start. Being first to market with differentiated solutions is a common goal among startup companies in order to attract funding from venture capitalists. While many startups fail, some have succeeded spectacularly and established a presence in new markets that, in turn, threatened the established companies in those markets. The significant advantage gained by being early to market with innovative solutions has not been forgotten by CEOs and senior business leaders at mature companies facing new competition.

Many of these same organizations also began to focus on managing their business by fact rather than by the intuition that drove their past decisions. Data became king and the information gleaned from it was deemed essential. The ability to look back in time and accurately assess what happened became a given. Using data to also predict the future became increasingly important when evaluating options and the potential impact of new decisions.

On the face of it, Big Data should help organizations respond to both of these needs. After all, more data variety and more data volume should help uncover new truths, or so many business executives would like to believe. And the Internet of Things seems to open up new business possibilities not only for strategies that can be used against traditional competitors but also for development of new strategies that can be applied in adjacent markets.

As a result, many IT organizations are tasked with coming up with a strategy to develop new solutions using Big Data in ways that will make a difference to the business. The Internet of Things is now receiving similar attention for many of the same reasons. A common initial approach to figuring out where such initiatives might provide value is to look for killer use cases uncovered at other companies in the same industry. Another approach is to simply try to make significant and unexpected business discoveries by exploring massive amounts of diverse data and hope that some "eureka" events occur. However, these approaches rarely work without a solid hypothesis as to the business problems that might be solved through the analysis of all data, including data coming from sensors, social media, web sites, and other streaming data sources.

29

In order to develop such hypotheses, a visioning session could be in order in your company or organization. Certainly, IT executives, enterprise architects, and IT architects will have a view on where IT is headed and also have some awareness of the potential impact of Big Data and the Internet of Things. However, the use cases being sought will most likely be in the minds of business leaders. Coincidentally, they might also have funding for the budget needed to pay for such projects. Individuals that have described their needs for such projects in planning sessions that we have facilitated include chiefs of marketing, heads of risk management, and vice presidents of engineering (just to name a few).

In this chapter, we describe how to discover what could be in your future information architecture and drive future projects by evaluating "the art of the possible." You might find many potential projects when this session occurs in your organization. But the techniques outlined in this chapter will also help you to develop a realistic early assessment as to how viable the desired projects are. As a result, you will be able to focus on the projects that really do have the right level of support and can make a difference to the business.

Figure 2-1 highlights the stage we are at in our methodology and what we are covering in this chapter. The discussion in the visioning session includes an evaluation of the current and future business architecture, data architecture, application architecture, and technology architecture. Since this is just the start of our discovery process, there will be many follow-up stages once we have established a vision and are convinced there might be a project worth pursuing.

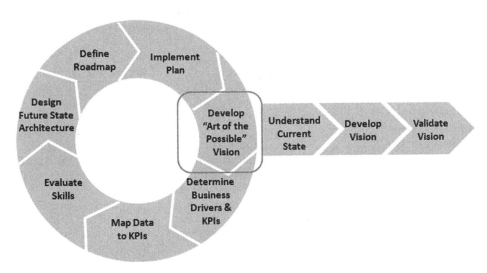

*Figure 2-1.* Art of the Possible Vison phase in our methodology for success

# Understanding the Current State

Before laying out a vision of where we might be going, it is important to understand where we are. There are always both business and technical views of the current state. From a business perspective, we need to understand if our business executives and analysts are satisfied with the information and data they have. This often leads to a discussion about how and why they use data, the data granularity, the breadth of history that is accessible, and the quality of the data. Missing sources of data and data history are discussed as well as the impact that adding this data will have on volume requirements in the future. The desire for timely data and what is acceptable timeliness is also discussed.

From an IT perspective, we need to understand current key data sources, how and where the data is moved, the data management systems that are utilized, and the business analyst tools currently used or lacking. In addition to the software, we should understand the capabilities and age of server and storage components in the current state architecture. We should also understand if service level agreements to the business are being met and how flexible and agile the technical infrastructure is when the business must respond to changing conditions.

## Information Architecture Maturity Self-Assessment

An early self-assessment of the maturity of the current information architecture can yield insight as to the ability of an organization to extend its current architecture. If an organization is struggling with a basic data warehouse implementation, there should be little expectation that taking on a Big Data project will magically fix all of the problems. In fact, such a project could get in the way of solving higher priority problems that the lines of business would prefer be solved sooner.

There are a variety of maturity rating scales for information architecture found in publications. We have found that organizations generally follow a path that can include starting with silos of information and data, then standardization of information and data, application of advanced business optimization techniques, and providing information as a service. Figure 2-2 illustrates this path.

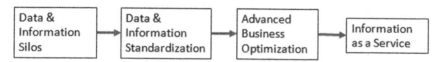

*Figure 2-2.* *Information architecture maturity stages*

Following are definitions of what happens in each of these stages:

- Silos of information and data: Data is duplicated inconsistently across many independent data marts and tools, primarily managed in the lines of business, and there are frequent debates about which data set is the true data set. As a result, the interpretation of any results coming from data analysis is often viewed with skepticism.

- Standardization of information and data: Centrally managed dependent data marts and enterprise data warehouse are used for reporting and ad hoc queries, with attention paid to data quality, consistency, and security. IT and the lines of business collaborate on data mart extensions and rollout of new marts.

- Advanced business optimization: Streaming data is introduced augmenting traditional data sources. Predictive analytics is used to better understand and predict outcomes of decisions.

- Information as a service: The internally developed trusted data stores and analysis tools have value outside of the company or organization. Access is provided to favored partners as a benefit of doing business. Subscribers are willing to pay for the service and can be provided access.

These stages are not always pursued in a sequential manner. Organizations may be traversing several of these initiatives at the same time. For example, organizations often gyrate between data silos and standardization, especially where IT doesn't move fast enough to meet changing analysis needs from the lines of business. Predictive analytics and the addition of streaming data are sometimes explored and implemented while this gyration is occurring.

Of course, organizations that successfully traverse these first three stages will have created something of incredible business value. At that point, some consider setting up subscription offerings and might go into competition with data aggregators in their industry as they begin to provide information as a service.

As you evaluate the maturity of your organization, it is important to realize that as you move from silos to information as a service, the role and skills that IT must bring to the organization become more advanced. Part of the consideration in taking on a new project should be whether you are introducing a significantly different skill set to your organization and whether the investment needed in gaining these skills should be spent here or on other less demanding but possibly equally business lucrative tasks.

## Current Business State of the Industry

An understanding of industry trends and how best-of-breed competitors are redefining their information architecture to address those trends is also important when developing the future information architecture. Keep in mind that the introduction of Big Data and the Internet of Things is leading to a redefinition of who the competition is in many industries. Some are choosing to enter other industries based on an ability to make sense of data in new ways, thus enabling new business entry points and solutions.

The most impactful information architecture projects are always linked to solving specific business problems. The following is a sample list by industry of typical data warehousing projects and projects where the information architecture is extended to include Hadoop and / or the Internet of Things. This list may give you a few ideas of areas to explore for new projects that could yield significant return on investment when aligned to business goals in your organization:

- Agriculture:
    - Data warehousing: Cost of farm production and optimization, yield analysis, agricultural goods commodity pricing / trading analysis.
    - Hadoop / Internet of Things: Analysis and optimization of plowing patterns, fertilization, readiness for harvesting, and moisture content (from sensors in the field and weather data).

- Automotive Manufacturing:
    - Data warehousing: Cost and quality of manufacturing, supply chain analysis, warranty analysis, sales and marketing analysis, human capital management.
    - Hadoop / Internet of Things: Analysis of customer sentiment and analysis of connected vehicles including component failure, need for service and service scheduling, driving history (and automated car), driver emergency detection and response.

- Banking:
    - Data warehousing: Single view of customer across financial offering channels, financial analysis, fraud detection, credit worthiness, human resource management, real estate management and optimization.
    - Hadoop / Internet of Things: Fraud detection, risk analysis, and customer sentiment.

- Communications:
    - Data warehousing: Pricing strategies and finances, customer support and service, marketing analysis, supply chain, logistics and process optimization, regulatory compliance, real estate optimization, and human capital management.
    - Hadoop / Internet of Things: Analysis of social data, mobile device usage, network quality and availability (using sensors), network fraud detection, and for Internet of Things, extended network management and optimization.

- Consumer Packaged Goods:

  - Data warehousing: Analysis of sales, marketing, suppliers, manufacturing, logistics, consumer trends, and risk.

  - Hadoop / Internet of Things: Analysis of promotional effectiveness (through social media and in-store sensors), supply chain, state of manufactured goods during transport, product placement in retail, and risk.

- Education and Research:

  - Data warehousing: Financial analysis of institution or facility, staffing and human capital management, and alumni profiling and donation patterns.

  - Hadoop / Internet of Things: Analysis of students at risk (using sensor data), research data from sensors, and facilities monitoring and utilization optimization.

- Healthcare Payers:

  - Data warehousing: Analysis of cost of care, quality of care, risk, and fraud.

  - Hadoop / Internet of Things: Analysis of sentiment of insured customers, risk, and fraud.

- Healthcare Providers:

  - Data warehousing: Analysis of cost of care, quality of care analysis, staffing and human resources, and risk.

  - Hadoop / Internet of Things: Disease and epidemic pattern research, patient monitoring, facilities monitoring and optimization, patient sentiment analysis, and risk analysis.

- High Tech and Industrial Manufacturing:

  - Data warehousing: Supplier and distributor analysis, logistics management, quality of manufacturing and warranty analysis.

  - Hadoop / Internet of Things: Shop-floor production and quality analysis, quality of sub-assembly analysis, product failure and pending failure analysis, and automated service requests.

- Insurance (Property and Casualty):

  - Data warehousing: Sales and marketing analysis, human resources analysis, and risk analysis.

  - Hadoop / Internet of Things: Customer sentiment analysis and risk analysis.

- Law Enforcement:
  - Data warehousing: Logistics optimization, crime statistics analysis, and human resources optimization.
  - Hadoop / Internet of Things: Threat analysis (from social media and video capture identification).

- Media and Entertainment:
  - Data warehousing: Analysis of viewer preferences, media channel popularity, advertising sales, and marketing promotions.
  - Hadoop / Internet of Things: Viewing habit analysis (from set-top boxes), analysis of customer behavior at entertainment venues, and customer sentiment analysis.

- Oil and Gas:
  - Data warehousing: Analysis of drilling exploration costs, potential exploration sites, production, human resources, and logistics optimization
  - Hadoop / Internet of Things: Drilling exploration sensor analysis (failure prevention)

- Pharmaceuticals:
  - Data warehousing: Clinical trials including drug interaction research, test subject outcome analysis, research and production financial analysis, sales and marketing analysis, and human resources analysis.
  - Hadoop / Internet of Things: Analysis of clinical research data from sensors, social behavior and disease tracking (from social media), and genomics research.

- Retail:
  - Data warehousing: Market basket analysis, sales analysis, supply chain optimization, real estate optimization, and logistics and distribution optimization.
  - Hadoop / Internet of Things: Omni-channel analysis and customer sentiment analysis.

- Transportation and Logistics:

  - Data warehousing: Equipment and crew logistics and routing, sales and marketing analysis, real estate optimization, and human resources analysis and optimization.

  - Hadoop / Internet of Things: Traffic optimization (from highway sensor data), traffic safety analysis and control, equipment performance and potential failure analysis (from on-board sensors), logistics management (from sensors), and customer sentiment analysis.

- Utilities:

  - Data warehousing: Logistics optimization, grid power delivery analysis and optimization, customer energy utilization, and human resources analysis and optimization.

  - Hadoop / Internet of Things: Analysis of data from smart meters for grid optimization and status, pro-active maintenance optimization.

---

■ **Note**    The preceding list is representative of just some of the projects and implementations that were in place or in progress in 2015. This list continues to change as organizations find new and innovative uses for the technology and seek new business solutions to previously unsolvable problems.

---

Later in this book, we will discuss prioritization of these kinds of projects. Prioritization (and often funding) requires that the lines of business work in partnership with IT. The odds of ultimate project success improve dramatically when project definition and prioritization is a joint activity.

## Is a New Vision Needed?

At this point, we have self-evaluated our information architecture maturity and vetted some possible projects. It might be possible to continue to modify our existing information architecture in minor ways in pursuit of desired new projects. For example, if the data needed for analysis is largely structured and the data warehouse infrastructure is sound, it might be most efficient to simply build upon that architecture. However, where the infrastructure and business needs are not aligned, now could be the right time to come up with a vision on how to address these growing needs.

An obvious reason to come up with a modified architecture (as you might have guessed from the topic of this book) is the identification of business needs that require analysis of new sources of data not easily handled in a data warehouse. For example, new data sources might include streaming data or semi-structured data. A high-speed and high-volume data ingestion requirement could be introduced. Such requirements can lead to the inclusion of NoSQL databases and Hadoop into the information architecture where they might not have existed or been necessary before. The data scientists exploring Hadoop might desire and drive the adoption of new information discovery tools and predictive analytics engines.

How do you gather needs and collaboratively develop a vision of the future state information architecture? Companies and organizations we have worked with typically hold planning sessions (sometimes called workshops) and gather requirements. Gathering the initial requirements might take only two or three hours, but it can help determine the direction for all that is to follow.

During the visioning session, attendees should discuss their level of satisfaction with current data warehousing, business intelligence, and ETL tools and processing solutions. This discussion might extend to underlying infrastructure including the servers and storage. It is often during this phase that upgrades or changes to existing components are first considered. For example, if ETL performance and resources required for ETL on the target data warehouse is a problem, it might make sense to consider leveraging a Hadoop cluster needed for streaming data sources to also become an ETL engine. In fact, the Hadoop engine could become the initial landing point for all data. If there is an emerging need for predictive analytics or a re-evaluation taking place as to how predictive analytics is deployed, you might explore the Hadoop cluster for handling that workload as well.

It is usually during this phase that many in IT will begin to worry if the maturity of the current information architecture is badly out of alignment with the vision that is being discussed. Skills gaps in IT or the lines of business might become all too apparent. Data governance and operational questions could also arise. Potential cost and impact on the budget are usually at the forefront of concerns in IT senior management.

All of these are good to call attention to, even at this stage. However, we are only at the beginning. We are building a vision of what the future information architecture might become. We don't yet understand enough about the business case (including whether we will have one) to be sure that we will be able to pursue a project. We also don't have enough detail about the data needed. We'll better understand all of that in later stages. Later, we'll also more closely evaluate the skills needed. At the point we start to define the more detailed future state information architecture, we'll begin to better understand the potential costs of the solution.

For now, we are simply exploring the art of the possible.

# Developing the Vision

The vision is often developed during a facilitated planning and whiteboard session. Even at this earliest stage of speculating about revising the information architecture, you will want a variety of key stakeholders in the session. Certainly, you'll want the architects and IT management that understand the current architecture and its components to be present. But at this point, your organization's business executives and analysts could have the clearest picture as to what new sources of data must be included in order to answer new and emerging business problems. The business analysts might also have a distinctly different view from IT as far as how granular the data needs to be and how long the data must remain. Remember that now is not the time for debate. This is the time to gather everyone's requirements.

Of course, it won't hurt for you to do your homework ahead of the session. If the CIO or other senior IT leadership is not in regular attendance at business planning meetings, you will want to review the organization's top business priorities as articulated in earnings calls and earnings statements, internal broadcasts, and other internal forums. You might do similar research on your competition since business executives at your company are likely keenly aware of competitive pressures.

---

■ **Note** In companies and organizations where IT has grown disconnected from the lines of business and is not seen as a trusted or reliable delivery partner, we have seen IT groups try to independently define and pursue information architecture modifications that they believe will be of general interest. Such efforts have a tendency to remain in research and development with little discretionary funding available until trust is re-established with the business and joint visioning and planning occurs.

---

To draw the right attendance to your session, you should circulate the visioning session goals and an agenda prior to the meeting. For example, a defined goal might be the desire to gain early input on a five-year plan for IT investment that will enable your company to provide better customer experience. Note that we've called out a business goal in addition to an IT goal. If we want business people to be in the meeting, we need to describe what is in it for them.

An agenda outlining what will be discussed at a visioning session could resemble the following:

- Overall goal of the session

- Self-introduction of attendees and self-described meeting goals

- Outline of the kind of information to be gathered

- Discussion of current information architecture maturity and its implications

- Review of current information architecture footprint and business solutions provided

- Discussion of what needs to change and why

- Vision of how the information architecture and business solutions could evolve

- Discussion of next steps and other sessions needed

You should make it clear at the outset of the session that input and the whiteboard drawings will be captured. You should also promise up front that, after the session, a report will be provided and distributed summarizing the information gathered.

# The Current State and Future State Data Warehouse

The IT group likely has developed highly technical and detailed current state information architecture diagrams. These can serve as useful reference materials as we start the process of determining the information architecture evolution that might occur. However, we'll want to simplify these diagrams and focus on the pieces relevant to the needs of the business areas taking part in formulating the vision.

Figure 2-3 represents how a current state footprint might be represented where the focus of the visioning exercise is on improving the success of promotions and sales efforts in a company that has retail stores. We would use this or a similar diagram as our starting point. In this example, the Enterprise Data Warehouse (EDW) provides the historic database of record. Data is extracted from multiple OLTP systems (the ERP and CRM systems are pictured). Data marts surround the EDW. Business analysts access the marts and / or EDW using reporting and ad hoc query and analysis tools. We've also indicated some of the key technologies in the figure that are part of the current footprint as we may want to reference them during the discussion.

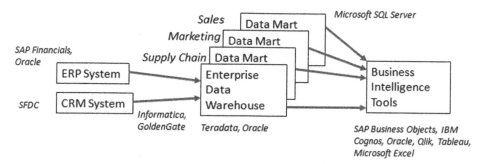

***Figure 2-3.*** *Typical current state information architecture diagram*

During the discussion about this architecture drawing, many topics can be explored in more detail. Some topics that often come up in this discussion include the following:

- Current granularity of data and length of history in the EDW (and what is desired)

- Current usefulness of data marts, including overlap between them, agility in extending or deploying new marts, and length of history needed

- Current usefulness of reports and business intelligence tools and dashboards

- Possible standardization of business intelligence tools if several exist with overlapping functionality and relative sizes of communities of users

- Need for dimensional models with conformed dimensions among data marts

- Need for predictive analytics and data mining

- Level of satisfaction regarding current query and analysis performance

- Need for real-time alerting or recommendations

- Importance of third-party data sources and applicability of that data to one or more data marts

- Need for additional sources of data to address emerging business problems

- Current quality of data in the EDW and data marts and desired future quality standards

- Frequency of data updates and ability to deliver data to the EDW or marts (and ultimately to the analysts) in time to make business decisions

- Concerns about data security including data at rest and in motion

- Need for a more highly available EDW or marts

Gather as much as you can from all present and make notations on the whiteboard as needed. You will want to document names given to data marts, sources, and other components so that everyone has a common understanding of what is being discussed.

---

■ **Note**   If you are only vaguely familiar with the current state information architecture and other architects present in the session wants to correct your drawing, gladly give them the marker and let them draw their own version. Their drawing might look considerably different from the illustrations we provide in this chapter, but seeing their version can be tremendously enlightening to all present regarding how the current architecture is viewed. This will build a better joint understanding of what exists and better collaboration regarding what might be changed as the vision of the future state architecture is determined.

---

As we just noted, data security can come up in this discussion. Some of the discussion could be driven by unique requirements of the industry that the organization is part of. If you work for a company in a specific industry or a government agency, you are probably familiar with standards that must be enforced. However, if you consult among many different industries, Table 2-1 can provide a useful description as to some of the more common standards often mentioned.

*Table 2-1.* *Sample of Data Security Standards in Various Industries*

| Industry | Standard Name | Description |
|---|---|---|
| European Union | European Union Data Protection Directive (EUDPD) | Requires all EU members to adopt security directives |
| Financial / Banking | (US) FFIEC security guidelines for auditors | Defines online banking security requirements |
| Financial / Banking & Services | Gramm-Leach-Bliley Act (GLB or GLBA) | Sets requirements for privacy and security of financial data collected |
| Financial & Retail—Payment Processing | Payment Card Industry Data Security Standard (PCI DSS) | Framework for payment card data security processing: authentication, fraud detection, and prevention |
| Healthcare—Medical Records | Health Information Portability & Accountability Act (HIPAA) | Access control, auditing, data integrity, encryption standards for healthcare data |
| US Department of Education | Family Educational Rights & Privacy Act (FERPA) | Privacy standards for student data including grades, enrollment, billing |
| US Government | Federal Information Processing Standard (FIPS) from NIST | Defines authentication, cryptographic key management, and physical security at US agencies |

You might also begin to explore with the business analysts the need for readily accessible sandboxes where they can explore data from a variety of sources on an as-needed basis. In the past, these sandboxes were sometimes made available in EDWs or as disposable data marts. More recently, information discovery tools have emerged that provide their own data management engine or use Hadoop for data management. These tools are more flexible and timely for exploration of new data because they are "schema-less"—in other words, a schema does not need to be pre-defined for them to be useful.

Figure 2-4 adds the notion of an information discovery tool to our current state data warehouse architecture. Understanding the usefulness of intelligence gained by exploring data with such a tool can help us understand how to better utilize the data in a Hadoop cluster, but using such a tool might also help drive the requirements for future build out of data marts and traditional business intelligence tools.

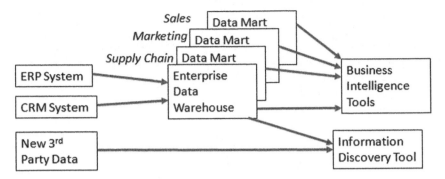

*Figure 2-4. Information discovery in the information architecture diagram*

Remember to gather from the business analysts present any tangible business benefits that they believe will come from the fulfillment of the reporting and analytics vision. They might be able to share data on industry best practices or have studied their own organization's efficiency and offer some compelling evidence of the value that could be provided. This information will be important later as we explore in more detail how to develop a business case for the project and prioritize it in comparison to competing projects.

## Determining Where Hadoop and NoSQL Databases Fit

We are now ready to probe for the potential need for extending the information architecture to include Hadoop and NoSQL databases. This extension will likely be driven by the need to handle streaming and semi-structured data sources more efficiently than is possible using relational database technology.

If your company is like many organizations, you could already have research and development efforts trying to determine the value such data might provide. If your organization already has such projects, you will want to document these as part of your current state. For example, you might explore the following topics if a Hadoop prototype is underway or if Hadoop is already part of your production environment:

- Current sources of data to the Hadoop cluster

- History length the data represents that is loaded into the Hadoop cluster and its current volume

- Desired additional data sources to be loaded into the Hadoop cluster

- Desired history length of data that will be stored and impact on data volume

- Current and future planned data ingestion rates

- Current and future planned workloads in the Hadoop cluster (MapReduce, SQL query, Solr / search, predictive analytics, ETL, and so forth)

- Current and future planned analysis tools (business intelligence, information discovery, search, predictive analytics, ETL, and so on)

- Level of satisfaction regarding current ability to make sense of the data and drive business value (including any challenges related to skills)

- Concerns about data security including data at rest and in motion

- Concerns about recoverability and availability of the data in the cluster

Figure 2-5 illustrates how many projects that analyze data in Hadoop are started. The figure represents a typical Hadoop initiative at a company that has retail stores. In it, the Hadoop-based effort is represented as entirely separated from a pre-existing data warehouse and its surrounding infrastructure.

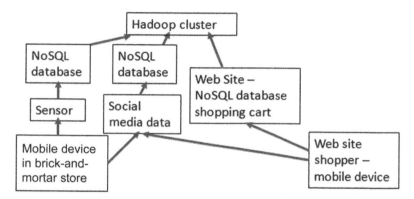

**Figure 2-5.** *Hadoop and NoSQL databases in a separate research and development effort*

In the theoretical retail store example, the business goal is to better understand the success of promotions and sales efforts. Data is gathered in Hadoop after being captured when shoppers buy items and browse on the web site and when they enter the brick-and-mortar stores. Sentiment data that expresses the shoppers' opinion of doing business with the company is gathered in Hadoop from social media. The streaming data landed in clusters of NoSQL databases that can easily be scaled for high-ingestion demands and then loaded into Hadoop for analysis.

Note that the gathering of data from sensors is represented in a very simplified view of the Internet of Things architecture. Much is missing from the diagram including provisioning, security, and other necessary services. For the purpose of defining a vision as to how all of this will fit together, this is fine for now.

# Linking Hadoop and the Data Warehouse Infrastructure

We will next determine if there is a need to query and analyze data residing in our traditional data warehouse information architecture and the Hadoop cluster at the same time. Understanding the sophistication of our analyst community and the frequency that they will access data from both sources in combination in order to answer business questions will help us determine the optimal footprint to recommend.

For example, if we need to provide dimensional models to enable business analysts to explore the data, we'll either need to move data of value from the Hadoop cluster into the data warehouse or provide an infrastructure that can use the Hadoop cluster as extended storage for the data warehouse. If we plan to provide a platform for predictive analytics that includes data from both, we likely will want all data to be analyzed to reside in the Hadoop cluster. As noted previously, we'll want to gather from the business analysts the potential business benefits of analyzing these various sources for data together.

Figure 2-6 illustrates how the Hadoop and NoSQL databases often become part of the existing data warehouse information architecture. Since we might also want to leverage Hadoop for ETL processing, we've pictured the data sources that formerly directly fed the enterprise data warehouse as becoming feeds to Hadoop. This, once again, is a highly simplified diagram.

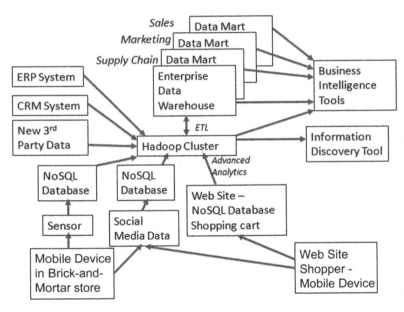

***Figure 2-6.*** *Hadoop and NoSQL databases linked to the data warehouse infrastructure*

At this point in a visioning session, there is sometimes a debate about where the data should be hosted. Should the data from these new data sources land first and be analyzed in the cloud? Should data land instead in an on-premise infrastructure that includes an existing data warehouse and the Hadoop cluster?

Data volumes and the amount of movement required over networks (with limited bandwidth) will help you decide whether the data might be most appropriate in the cloud, on-premise, or in a hybrid model. In order to understand the amount of data movement, you'll need to understand the query and analysis activity against data in the various data management systems and their locations. These options should be explored in more detail later as the technical information architecture is better defined. At that point, you'll have a much better idea as to the business use cases envisioned and the data needed to solve those.

---

■ **Note**    Many organizations use the cloud for rapid deployment of Hadoop, NoSQL, or data warehouse research and development efforts today, especially where business value is unknown, and then bring the infrastructure in-house when they consider the growing data volume and customization requirements of the overall production infrastructure.

---

# Real-Time Recommendations and Actions

It should be evident that the infrastructure pictured in Figure 2-6 introduces significant time delays (in other words, latency) due to the movement of data from point to point. For certain problems the business might encounter, there could be a need for real-time recommendations and actions in response. For example, when a shopper is choosing the products to buy on a web site, you might want to intelligently recommend other products while they are shopping on the site, not some time after they have left the site.

For such scenarios, a real-time recommendation engine is used to guide the shopper by presenting specific products in the web store while they are engaged. Predictive analytical models of buying and shopping behavior are processed in the Hadoop cluster or data warehouse. Of course, the goal is to make smart recommendations such that the shoppers will buy more and find what they are looking for faster. The models in the recommendation engine are periodically updated with current buying patterns and the engine appears to become smarter in the recommendations it makes over time.

Where intelligent sensors and controllers are deployed, a critical need for timely action might suggest that certain rules be established to drive action before any analysis occurs. That is why event processing and business rules engines are often deployed as part of intelligent sensor solutions today. For example, if sensors in the brick-and-mortar store begin to detect delays in reaching cashiers and dissatisfied customers abandoning the items they wanted to buy, predefined rules might trigger devices to signal cashiers who are engaged in other activities to open up additional cash registers and alleviate the backup.

Figure 2-7 illustrates the addition of a real-time recommendation engine to our web site with data models run in the Hadoop cluster and periodic updates fed to the engine. Shopper profile and location information is passed to the engine. Specific recommendations are passed through the web site back to the online shopper.

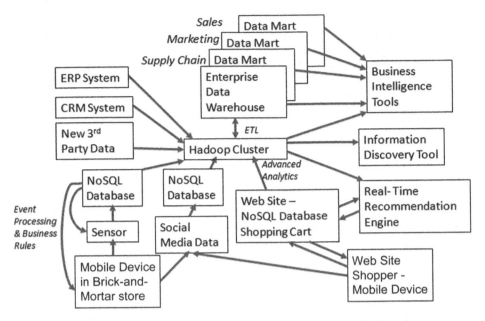

**Figure 2-7.** *Real-time recommendation engine and event processing in the information architecture*

We also illustrated the addition of closed-loop event processing and business rules where the sensors are deployed in the brick-and-mortar stores. In addition to our cash register bottleneck example, as shoppers enter the store with mobile phones that run our loyalty application, we might want to begin monitoring where they are located and have our salespeople better assist them based on information we have gathered on products they have recently been shopping for on our web site.

The diagrams we just presented might seem a bit technical for business analysts to comprehend. However, our experience is that such diagrams can help business analysts understand the limitations of the current infrastructure and the data flow and analyses options available to solve their problems. The diagrams also help to get early buy-in on what needs to change in order to deliver the business solutions that they envision.

# Validating the Vision

The visioning planning sessions usually end with the current state and future state architecture diagrams drawn on a whiteboard, and these are often captured by taking photos of them using mobile phones. Other times, the diagrams are retained on tear-off sheets from a flip chart. A lot of notes should also have been gathered by the facilitators of the session during discussions about the current state and desired future state. The notes should include the input from the lines of business, senior IT management, the architects, and anyone else present. Some of the notes likely include assumptions as to the impact that the future state information architecture will have on the ability to make better business decisions once the architecture is deployed.

It is now time to summarize all of the information that was gathered in a more formal way. That information should be delivered back to all interested parties that took part in the visioning session in the form of a report or presentation. Doing so will help validate whether all of the vital information about future requirements was captured. It also will likely spur some further clarification as to what is needed. Generally, providing the information back to the original attendees within a week or two is a best practice and helps maintain the teamwork that was established in the original session.

---

■ **Note**   In some engagements, we've received requests for a substantially different audience to be invited to the follow-up session, a session that was intended to summarize what was discovered in the earlier planning session. This sometimes occurred because the original attendees were excited about the potential future state architecture that was discussed and shared that excitement, and then additional stakeholders became interested. In other situations, leaders from other lines of business became interested in the process and wanted to make sure their needs were documented as well. This is a positive indication that many in the organization believe that such a project would be of value and could be funded and implemented in the future. Keep in mind that if the follow-up session becomes more focused on further discovery than on the read-out and verification of the earlier work, a good approach is to offer to create a further revision of the report and then schedule another read-out and verification of that revision.

---

The content that should be included in the read-out report or presentation of what was uncovered during the visioning planning session usually contains the following:

- Current business challenges including those created by the current information architecture

- Current state information architecture description and diagram

- Emerging business needs, likely further business model changes expected, and what will be needed to run the business effectively

- A future state information architecture diagram that could address the needs and challenges

- Business benefits that might be achieved with future state information architecture in place (ideally including the possible financial scale of those benefits)

- Next steps including scheduling of activities outlined in subsequent chapters of this book

In some organizations, we have observed that IT architects would like to take the information gathered during this visioning activity and immediately begin a detailed design of the future state information architecture. However, there is still a lot that is unknown. We don't yet have details on critical success factors tied to running the business, including the key performance indicators and measures needed. We don't yet have an idea as to how we might implement the desired solution in project phases or what priority the phases would have relative to each other.

Because of this, we don't yet have a solid business case, though we might have some initial ideas as to where the business case will come from. Our knowledge of which data sources will provide the needed measures and map to KPIs is extremely limited. Furthermore, we are not sure what impact the (lack of) skills in our organization could have on our ability to deploy and manage the envisioned solution or take advantage of it to run the business more efficiently.

At this point, we still need more discovery and further documentation before we will have a project clearly defined. But at least we now have some notion of the practicality and likelihood of modifying our information architecture, and we have a better idea as to where to look for justification. Furthermore, we are beginning to get an idea as to where business sponsorship might come from. Thus, we will next work with our lines of business partners to further uncover what is needed.

# CHAPTER 3

■ ■ ■

# Understanding the Business

In this information-driven age, we have reached the crossroads of a data revolution. We are at an inflection point in using revolutionary technologies provided by Big Data and Internet of Things (IoT) to gain intelligence, streamline operations, and gain competitive business advantage. Augmenting and transforming existing data warehouses and analytics footprints is the order of the day at many organizations. Before embarking on a journey to incorporate newer technologies, we should ask ourselves few basic questions and answer "why" we were doing this and "how" the technologies can support the business strategy. Even though these questions sound intuitive and basic, it is the right place to start. Our experience is that organizations that understand business drivers, critical success factors, and priorities before embarking on the journey are far more successful. This also helps when garnering internal support needed to launch the initiatives faster and effectively, while successfully reaping the projected business benefits.

In this chapter, we describe how to identify and analyze the business challenges and needs of an organization before formulating the future state information architecture. Figure 3-1 highlights the phase we are in. In this phase, we begin to uncover in more detail the business architecture required. Later in this book, we will embark on a journey through the technical architecture phases.

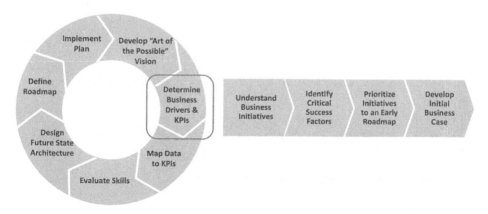

***Figure 3-1.*** *Business drivers and key performance indicators (KPIs) in our method to success*

# Understand Business Initiatives

Linking business needs and innovation to our information architecture initiatives will help us accelerate business value realization. We will also be able to plan more effectively. The growth and variation in data sources and changes in information demand patterns create challenges for anticipating and prioritizing information needs across lines of business. Typically, the innovation in information architecture that utilizes newer technologies can be hindered by an inability to align new analytic capabilities with desired business outcomes and by a lack of prioritization of information needs. Organizations should formulate their information strategy to meet data and analytics needs with a clear focus on business outcomes. Aligning information delivery to business outcomes improves IT responsiveness and enables organizations to direct investments to where they are needed the most.

## Big Data and IoT Impact on Business

The business intelligence and analytics capabilities that IT provides sometimes do not reflect added value to the organization. As shown in Figure 3-2, it is important to ask the right questions when trying to connect the dots from business strategy to IT systems, including how business strategy is shaping business goals and what information capabilities are needed in support of that strategy. We must understand where critical data sources are located and what potential data gaps there are. Within the scope and focus of this book, we seek to understand where IT solutions might need to include Big Data and IoT components.

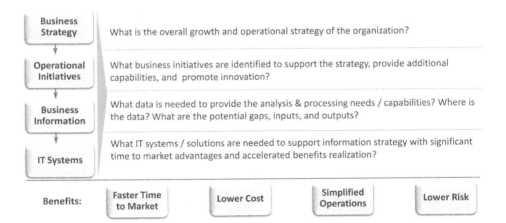

*Figure 3-2. Aligned priorities for higher business value*

Typical business intelligence and analytics applications focus on analyzing historic data and try to answer questions on how the organization performed in the past. However, this capability might not enable business analysts to react with agility and make business course corrections quickly. It is similar to looking into the rearview mirror to understand the path you have traveled. Even though operational analytics are very important in order to learn more about past business performance, adding predictive analytics can enable the analysts to see ahead, react quickly, and influence the organization's course and, thus providing higher business value. With the high volume, variety, and velocity of data typically attributed to Big Data and IoT initiatives, both predictive and operational analytics capabilities are usually provided in the unified information enterprise architecture. This information architecture approach can help organizations leapfrog to another level of business efficiency.

■ **Note**   Typically innovation in Big Data projects is hindered by an organizational inability to align new analytic capabilities with clear business outcomes and by an inability to prioritize information needs.

There are often just a handful of new technical capabilities that will add value to the business, as illustrated in Figure 3-3. It is important to understand how these capabilities can truly impact the business. The ability to determine the degree of their impact and to measure that impact enables organizations to plan and execute better.

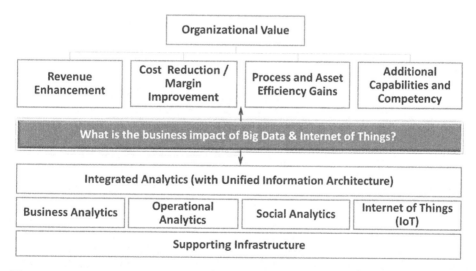

***Figure 3-3.*** *Representative business value drivers to align to anlytics needs*

Some of the typical impacts that might be measured and improvements that might occur include the following:

- Higher revenue: Measured by increased market share and penetration into new markets.
  - Increased volume of sales by building richer customer profiles, attraction of new customers, and cross-sell / up-sell while promoting innovation.
  - Ability to enter / create new markets or provide new services across the value chain for improved revenue.
- Enhanced margins: Measured by a reduction in direct and indirect costs of goods and services and higher returns from marketing investments.
  - Reduced cost and waste in the supply chains and agility to respond to increased demand generated by marketing campaigns.
  - Better customer retention and satisfaction through improved pricing strategies and enhanced demand and supply management.
- Improved efficiency: Measured by improved equipment and systems maintenance costs and availability.
  - Extended life, better availability, and improved efficiency of existing equipment.
  - Enhanced utilization through improved demand management.
- Competency: Measured by positive public and partner perception of the company and its ability to respond to external factors with agility.
  - Improved stock price and value of the company, better partner relationships, and faster response to competitive threats.
  - Significantly improved time to market capabilities.

---

▨ **Note** When Big Data and IoT initiatives are driven by IT alone and focused only on technical needs and current data, organizations often end up with a set of analytics platforms that provide little impact instead of platforms that deliver unified analytics solutions with much higher overall business impact.

---

Big Data and IoT technologies can provide transformative capabilities in the areas we just described. As a next step, we will need to understand how they might augment our information architecture and deliver these business benefits. Figure 3-4 illustrates a sample high-level strategy map. Many variations of this type of diagram are possible and such diagrams can be used to help align IT initiatives to business strategy.

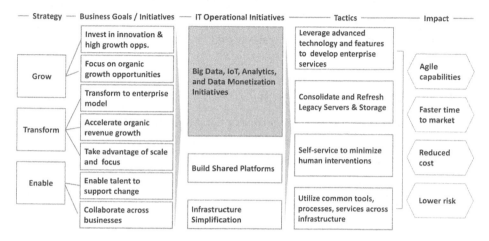

*Figure 3-4. Representative chart to align IT and information strategy to business strategy*

# Data Gathering Methods

Understanding business needs and aligning IT initiatives to business goals and strategy is critical to success in projects of this type. For many of these initiatives, the effort can be divided into three steps: plan discovery, preliminary research, and interviews with key stakeholders. We illustrate this process in the Figure 3-5.

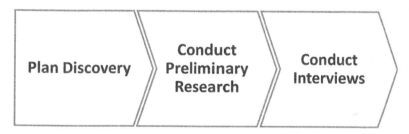

*Figure 3-5. Major components of discovery*

# Plan Discovery

Investing time in planning for discovery upfront will speed collecting the business drivers, challenges, capabilities, and priorities in a most effective and timely fashion. The result will be a better understanding of all of these. Any major planning activity should be defined by the following:

- Scope, objectives, and estimated timeline

- Information needed, potential sources of information, and data collection approaches

# Conduct Preliminary Research

Prior to conducting interviews, we should do our homework. Some of the possible sources of valuable insight include the following:

- Annual and quarterly reports

- Internal business strategy documents

- Key success metrics as reported by lines of business

- Industry databases and any other offline relevant sources of information

- Key stakeholder's individual performance measurements and compensation influencers

Preliminary research is typically a source of significant information that can help us formulate an initial value hypothesis and guide early lines of questioning during interviews. Generally speaking, the wider the variety of research, the richer the initial analysis will be. Even though high-value data is increasingly being made available online, the best information sources are often obtained offline. Persistence and creativity are always helpful in gathering as much information as possible to understand potential business drivers and use cases in support of Big Data and IoT initiatives.

# Conduct Interviews

At this stage of discovery, the focus is on gathering additional information and insight that is specific to the business use cases that are under consideration. The intent during this phase is to collect information that is exclusive, unpublished, and unique. Discovery through interviews and focus groups are common primary research methods and can produce qualitative and quantitative data. Here are a few key sample groups / resources that could be interviewed:

- Lines of business leaders (such as heads of marketing, supply chain, and so on)

- CFOs and the financial support organizations

- Corporate strategy groups
- CIO, IT leaders, and other enterprise and information architects in the organization

The extent of the interviews depends on the number of use cases being explored and the information that is needed. Many interviews can be used for collecting both qualitative and quantitative data from the same individuals. Collecting quantitative data requires a more structured approach and usually includes a combination of questionnaires and data collection templates. Face-to-face interviews are recommended when possible to collect effective information and to gain an understanding of business needs.

Group workshops and discovery sessions typically are designed to gather qualitative information. These sessions enable participants to reflect their thoughts on the use cases under consideration. The interactions can generate new ideas, and the exchanges can be very effective. One can get a sense of the group's opinion by comparing their responses to responses to similar questions provided by other key groups during discovery.

# Identify Critical Success Factors

For any information architecture initiative, a set of business and IT success factors should be identified. These factors are mapped to a project that will define a proposed future state architecture. The success factors typically resolve existing challenges and / or provide new capabilities that can drive business innovation and differentiation.

## Business Drivers

Business drivers influence the current and future business performance of an organization. It is important to understand the key business drivers for an organization and the industry that an organization is part of. We must understand how these drivers will influence our proposed future state. For example, Figure 3-6 has few representative key business drivers for the transportation industry.

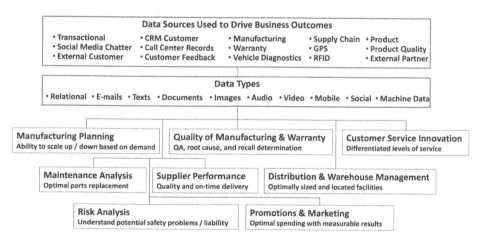

***Figure 3-6.*** *Future state drivers often under consideration in transportation companies*

Unlocking the power of data now often depends on how effectively organizations can combine structured, semi-structured, and streaming data at an enterprise level and develop a unified information architecture that aligns to business needs. The authors believe that analyzing data from IoT and other Big Data sources can lead to improved business outcomes. Table 3-1 provides a representative list of typical use cases in the transportation industry that can have positive impact on the business for companies in that industry.

***Table 3-1.*** *Representative Areas of Big Data and IoT focus for the Transportation Industry*

| Business Category | Areas of Focus | Benefits |
|---|---|---|
| Connected Modes of Transportation | Improved safety & security | Utilize vehicle sensor data for early fault detection, service alerts, and monitor emissions. Driver behavior / patterns monitoring for risk analysis and improved guidance. Sensor data analysis to prevent accidents / automated breaking. |
| | Additional location based services | Location-sensitive playlists by mashing podcasts, music, or news. Develop personalized advertisement campaigns and alerts based on vehicle location. |
| | Optimized navigation & improved driver experience | Smart route guidance and navigation based on real-time traffic patterns. Cabin personalization for operators. |
| | Reduced warranty costs | Real-time monitoring and remote diagnostics on ongoing basis will have direct cost-avoidance / savings potential in warranty claims. |
| | Insurance telemetric and new data monetization opportunities | Reduce claims-to-premiums ratio by analyzing risk profile based on real-time driving statistics. |

*(continued)*

*Table 3-1.* (*continued*)

| Business Category | Areas of Focus | Benefits |
|---|---|---|
| Connected Transportation & Enterprise | Accurate business demand planning & logistics | Improved warehouse and fleet management. Optimized demand & production management. Optimized inventory levels and improved logistics for predictable deliveries of goods and services. Enhanced partner relationships and integration for improved operations across the supply chain. |
| | Optimized operations | Fleet management and warranty efficiencies. Differentiated services and avoidance of costly public recalls. |
| | Higher business value and improved customer experience | Increase service revenue. Ability to launch new and improved personalized and mobility services for higher revenues. Overall cost optimization, better margins / improved price realization. Effective and personalized marketing campaigns, supply chain optimization, and pro-active customer engagement. |

Each industry and each organization has unique business use cases. We provided broad guidance on some of the likely use cases in Chapter 2. It is important to identify these during the discovery process.

Upon identifying the business use cases, identification of specific success factors in support of the use cases is crucial. We must identify critical success factors, key performance indicators, and key measures required. For example, if an organization wants to achieve 10% sales growth for their online sales, we should understand how sales are measured using the current state information architecture. We also want to understand how we can measure the impact on sales by the proposed future state architecture. As you will see later in this chapter, these are very important to understand when building a business case that includes quantification of projected benefits.

# IT Drivers and Linkage to Business Initiatives

Even though uncovering business initiatives are a primary focus for this section of the chapter, it is important to also discuss the IT drivers here. The alignment between the two is critical for developing an enterprise level and unified solution that effectively meets the business needs.

Once the business use cases that define the initiatives are identified, a technical evaluation of the environment and a gap analysis is needed to determine the current state vs. needed future state in support of the use cases. We introduced the concept of an information architecture maturity self-assessment in Chapter 2. On the maturity spectrum, Figure 3-7 also shows a linkage to organizational transformation. The authors believe that the majority of the organizations are now in the process of moving from a standardized environment to more agile and flexible environments. They are considering the benefits of analyzing a variety of data types. As the environment gets more advanced, the need for real-time actions and real-time recommendations becomes a desired and also feasible goal. Many of these thoughts are front and center in minds of IT departments today.

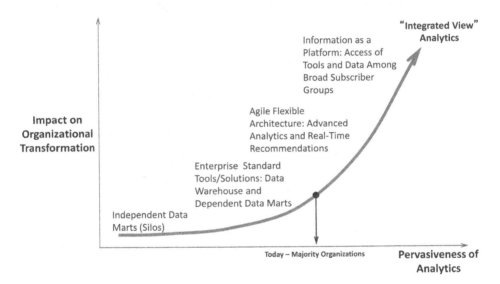

*Figure 3-7. Integrated analytics maturity continuiuum*

Well-planned initiatives consider both the business drivers and IT capabilities. Sometimes, a discussion of technical capabilities will open further discussion of possibilities on the business side.

A gap analysis can enable organizations to ensure that a proposed solution will provide the needed capabilities in support of the business needs. The following is a list of key attributes linked to major areas of consideration that can drive the need for a change in the information architecture:

- Strategy: Attributes can include overall strategy, capital and operational budgets, performance metrics, sponsorship, and project and program justification.

- Technology: Attributes might describe the appropriateness, applicability, integration, and support for standards, as well as the performance of technology and the IT architecture for the relevant workloads.

- Data: Attributes can include the quality, relevance, availability, reliability, governance, security, and accessibility of data of all types.

- People: Attributes that might be considered include technology and analytics skills, intra- and intergroup collaboration, as well as organizational structures, leadership, training, and cultural readiness.

- Process: Attributes can include data collection processing, data consolidation, data integration, data analysis, information dissemination and consumption, and decision making.

The technology and data areas are particularly interesting due to a tight linkage between them that can enable fulfillment of business needs defined in the business strategy and business processes. As the infrastructure becomes more flexible, we are offered more possibilities for solutions. At this point, we will defer considering details regarding people and organizational skills to Chapter 5 in this book.

---

■ **Tip**  Ensure that the technical solution selected is evaluated on all the major attributes of the information maturity scale. We describe how to do this at various places in the book. This is critically important for higher benefits realization.

---

Figure 3-8 illustrates some of the key technology components and data attributes that help define the unified information architecture. At this point, we have a vision as to how the future state information architecture might incorporate some of these attributes. As we build an early business case to better understand whether a project or projects might be remotely feasible, we'll begin to understand the potential scope of the effort and potential costs.

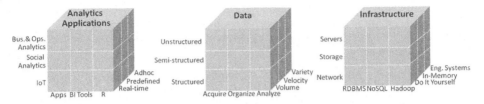

***Figure 3-8.*** *Key components that often define an information architecture*

Identification of technical and business drivers will enable organizations to select the right solution and start real project planning sooner, while helping to quantify the value to IT and the business.

# Prioritize Initiatives to an Early Roadmap

When we embark on a transformational journey, an evolutionary approach is more likely to be successful than a big-bang replacement. This holds true for all information architecture projects including Big Data and IoT projects. Since some of these technology footprints are relatively new in most organizations, they are often deployed on an experimental basis, possibly starting as a proof of concept in a lab.

Moving beyond the experimental phase requires an understanding of potential business outcomes that the technology can drive and whether the projects will be impactful and transformative. If that is the case and multiple projects are being considered, phases are usually determined based on the importance of specific business outcomes and potential funding that might exist. Project funding priorities are usually determined by the potential return on investment that will be achieved by deploying the solution. The project phases with the highest return on investment that fall within the budget get funded.

You might envision a three-phase approach from identification of initiatives to determination of business impact and finally prioritization of initiatives, as shown in the Figure 3-9. As we have already spent much of this chapter describing how to uncover initiatives, we next look at determining business impact.

***Figure 3-9.*** *Prioritization approach*

# Determine Business Impact and Prioritize Initiatives

A consistent and systematic approach should be followed to link possible projects to business goals and to determine the degree of influence a project might have. As shown in Figure 3-10, business and process performance drivers and their outcomes are mapped to success measures that, in turn, are mapped to project initiatives. This approach assures that we will capture the potential return on investment to the business from our projects.

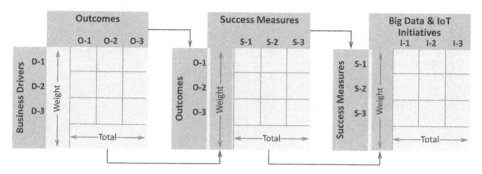

***Figure 3-10.*** *Prioritization approach*

Let's take a closer look at how we execute this approach in three steps.

Step 1: During this step, we identify top organizational goals and expected outcomes from them. Next, we assign a weight to the goals based on how important they are to the organization. For simplicity, we might use low, medium, and high scores (scale of 1 to 3). If teams prefer higher granularity, a scale of 1 to 10 can be used to assign scores. Similarly, we will indicate how the outcomes are believed to influence the goals. A scale of 1 to 3 or a scale of 1to 10 can be used to assign a score. It is very important to obtain stakeholder consensus at this stage to ensure complete buy-in. Once the scoring is complete, each outcome will have a weighted score representing how it will impact the organizational goals. (This score is obtained by multiplying each business driver weight with the score for the outcome and summing them up). Figure 3-11 provides a sample scoring card as part of prioritization in Step 1.

| | Low : 1 Med: 2 High: 3 | Weight | Outcome 1 | Outcome 2 | Outcome 3 |
|---|---|---|---|---|---|
| | | | **Outcomes** | | |
| **Business Drivers** | Driver 1 | 2 | 2 | 2 | 2 |
| | Driver 2 | 3 | 3 | 0 | 2 |
| | Driver 3 | 1 | 1 | 2 | 2 |
| | Total | | 14 | 7 | 6 |

*Figure 3-11. Step 1 of prioritization approach*

Some representative outcomes linked to effective warranty management in a manufacturing / automotive organization from various points of view might include the following:

- Financial: Ability to manage warranty costs by product.

- Customer: Ability to offer warranties that are competitive for premium products.

- Internal: Ability to link service and warranty fulfillment to determine if fraud is occurring.

- Innovation and Growth: Ability to use the warranty as part of product differentiation.

Step 2: For this step, we take the total scores from Step 1 and enter them as the relative weight for each outcome. Similar to Step 1, we score how each success measure can impact outcomes and calculate the score for each success measure. Figure 3-12 provides a representative calculation to come up with the scores for identified success measures.

| | Low : 1<br>Med: 2<br>High: 3 | Weight | Measure 1 | Measure 2 | Measure 3 |
|---|---|---|---|---|---|
| **Success Measures** | | | | | |
| Outcome 1 | | 14 | 2 | 1 | 0 |
| Outcome 2 | | 7 | 0 | 3 | 3 |
| Outcome 3 | | 6 | 0 | 2 | 2 |
| Total | | | 28 | 47 | 33 |

*Figure 3-12. Step 2 of prioritization approach*

- Continuing our example, here are a few representative success measures for the identified outcomes from the same points of view:

  - Financial: Reduce warranty reserve fund and save millions of dollars, Euros, and so forth. (Measure: Warranty cost by product)

  - Customer: Matched warranty offering helps improve product perception. (Measure: Product satisfaction)

  - Internal: Reduce reporting time reduced from five days to on-demand. A leading indicator into service is often visibility into warranty operations. (Measures: Defects linked to orders and number of product replacement orders)

  - Innovation and Growth: Increase sales by offering longer and profitable warranties. (Measure: Number of warranty claims per product over time periods)

Step 3: Next, we take the summary scores from Step 2 and use them as relative weights for success measures. We assign scores to each identified project initiative based on how they are expected to impact identified success measures. After scoring is completed, each initiative will have a final score to indicate how it is expected to influence the business outcomes. Initiatives with higher scores are expected to have higher business impact. Figure 3-13 highlights the scores arrived at for all the identified initiatives for a representative scenario.

| Big Data, IoT, and Analytics Initiatives | | | | | |
|---|---|---|---|---|---|
| Low : 1<br>Med: 2<br>High: 3 | Weight | Initiative 1 | Initiative 2 | Initiative 3 |
| Measure 1 | 28 | 3 | 1 | 0 |
| Measure 2 | 47 | 2 | 1 | 3 |
| Measure 3 | 33 | 0 | 2 | 2 |
| Total | | 178 | 141 | 207 |

*Figure 3-13. Step 3 of prioritization approach*

- A couple of examples of initiatives that might appear in our manufacturing organization example include the following:

  - By building a 360-degree view of the customer, we can increase service revenue by launching new and improved personalized and mobility services.

  - By performing remote diagnostics on an ongoing basis, we can provide better preventive and pro-active maintenance of vehicles resulting in direct cost-avoidance / savings in warranty claims.

---

■ **Tip** Obtaining initial consensus on success measures and the impact assessment approach usually gets all of the stakeholders on the same page. Ensure that all of the major stakeholders have an opportunity to vote in this process to help assure buy-in. Having many stakeholders provide scores during the impact assessment process will help make the process more acceptable and the scores more accurate.

---

## Other Prioritization Considerations

Once the initiatives have been scored and their impact to the business has been quantified, we need an early estimate as to the level of effort and number of resources required for each initiative. We also will begin to consider the cost, degree of implementation difficulty, likely time required to implement, and any other factors of importance to operationalizing the initiatives.

Although several initiatives could be identified across the lines of business, these additional factors will help us further assess the feasibility of the initiatives. Figure 3-14 is a sample chart that plots the initiatives on a grid that compares strategic impact and value to risk and degree of complexity. Such a diagram can help us define priorities and phases. Other axes might be selected and displayed in such a chart—the axes selected depend on what the company or organization values.

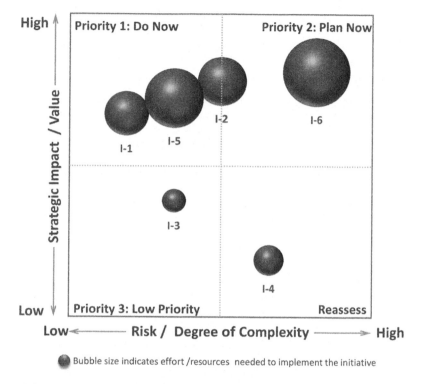

*Figure 3-14.* *Strategic roadmap*

We might also want to indicate other factors in our decision. For example, different colored bubbles might be used to identify a theme of transformation or to which line of business the initiative belongs to. Additional circles can be used around the bubble to indicate the time it takes to implement the initiative. There are many possibilities. The idea is to ensure that an informed investment decision can be made by using these charts, and we also use them in planning project phases.

# Develop Initial Business Case

Having a business case in support of a viable technical solution is important when moving key initiatives forward. It helps to justify the investments and also helps to garner the internal support for the initiatives. At this point, we don't have all of the information needed to do a comprehensive business case. We don't yet have a detailed technical design nor can we formulate in detail what the implementation might cost. But now is a good time to start to determine if our envisioned projects and future state architecture could be justified by a business case.

At this stage, we will focus on these three areas in order to build a simple business case:

- Total Cost of Ownership (TCO): The direct costs for solutions over a defined time period.

- IT Value: A quantification of IT process improvements and cost avoidance.

- Business Value: Business benefits provided by specific initiatives and their use cases.

We have observed that projects are most successful in getting off the ground when both IT and the business are engaged in developing the business case. These combined efforts can project the true total benefits of a new initiative.

As illustrated in Figure 3-15, quantifying benefits does get increasingly difficult and time-consuming when organizations move beyond simply computing TCO and begin to quantify projected IT and business benefits. This complexity occurs because of the involvement of various groups from lines of business, IT, and corporate, and difficulty in gathering high-quality data points. When key performance indicators are identified upfront, quantifying related benefits becomes easier and also accuracy of the business case improves.

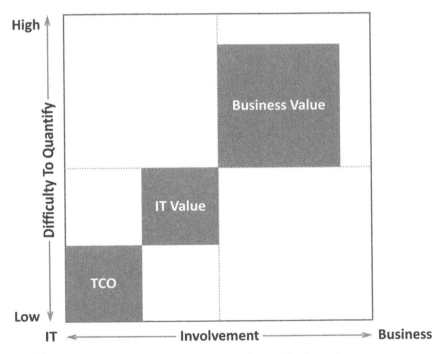

*Figure 3-15. Major componenets of a comprehensive business case*

---

■ **Note** When developing a comprehensive business case, one should focus on all three major areas of importance—TCO, IT Value, and Business Value. This will enhance the quality and completeness of the business case on hand.

---

The business case can help organizations evaluate various alternative options for a solution. Even though key technical aspects, solution implementation, and integration viability play major roles in short-listing the potential options, a financial business case can tip the balance toward an option and help drive the final decision while validating the need for the investment. The business case can help determine the degree that capital expenditures (CapEx) and operational expenditures (OpEx) fund the initiative, or how CapEx might be converted to OpEx should that be desired by the organization due to financing requirements.

When embarking on business case development, it is important to identify where all or portions of the funding will come from. For example, funding could originate in the CIO's organization, CFO's organization, lines of business, or from a corporate pool. There might be a chargeback mechanism in the organization or a central pool that could be looked upon as a cost center such that no chargebacks are applicable. If decision makers have their performance or compensation tied to business outcomes that the planned in-scope initiatives address, we would want to adjust our "language of value" when discussing these with the executives. It is important to ensure that all of these factors are considered when developing and presenting a business case to the decision makers.

Identifying how money is accounted for in an organization is another important aspect when formulating a business case. This aspect can vary and is commonly expressed in the language used by decision makers. Understanding the subtle aspects of their language enables analysis of the financials in a way that aligns with the organization. For example, some organizations will emphasize a run rate analysis over three to five years as they want to focus on OpEx reduction, while others focus equally on CapEx and OpEx.

While many organizations already have investments in labs or proof of concept environments to test next-generation information architecture components such as Hadoop, some have deployed next-generation footprints into production environments. In situations where organizations have already made such investments, it is important to find the expected growth rates needed to support the business initiatives, the age and depreciation of the existing equipment, and whether the equipment is owned or leased. These factors could also have an impact on the over-arching case for transformative investments in a revised information architecture that includes Big Data and IoT technologies.

# Total Cost of Ownership (TCO)

TCO computations should consider servers, storage, software, system environmental costs, installation, and implementation costs. All hard costs related to the solution options should be included for review, including any additional people-related investments that would have an impact on the direct costs. Here is a representative list of factors that might be considered for a simple TCO comparison of alternative solutions:

- Hardware:
  - Acquisition cost for nodes (servers and storage), sensors, and networking components
  - Annual support costs
- Software:
  - Software licensing cost (data discovery tools, business intelligence tools, data management software, application licensing, integration tools, and any other applicable components to support a comprehensive solution)
  - Annual support costs
  - Elimination of any software costs when software is replaced with newer technology
- Environmental considerations:
  - Power and cooling costs (We strongly recommend including these costs irrespective of the funding sources and any chargebacks in place for IT.)
  - Data center space costs (This could be a very big factor for data centers constrained by space limitations.)
- Installation and implementation:
  - Best initial guess regarding installation and implementation of the solution costs
  - Training needs and associated costs for the solution
  - Additional integration costs to other systems at the enterprise level

Typically we compare stand-alone solution TCO to a set of projected IT and business benefits. Table 3-2 provides an example of a high-level TCO summary for an 18-node Hadoop cluster. A similar TCO summary should be developed for each solution option under consideration for comparative purposes.

*Table 3-2. Sample TCO Summary Portion of a Business Case*

| Category | 5-Year Total |
|---|---|
| Hardware Acquisition | $327,330+$0+$0+$0+$0 = $327,330 |
| Hardware Support | $45,826+$45,826+$45,826+$45,826+$45,826 = $229,130 |
| Software Acquisition | $113,400+$113,400+$113,400+$113,400+$113,400 = $567,000 |
| Software Support | $0+$0+$0+$0+$0 = $0 |
| Floor Space, Power, and Cooling | $16,367+$16,367+$16,367+$16,367+$16,367 = $81,835 |
| Implementation, Migration, and Training | $62,400+$0+$0+$0+$0 = $62,400 |
| Total | $565,323+$175,593+$175,593+$175,593+$175,593 = $1,267,695 |
| Net Present Value (NPV) | $1,131,397 |

# IT Value

This portion of the business case involves quantification of IT process improvements and cost avoidance achieved. Some representative IT benefits might include the following:

- Process Improvements:

    - Ability to respond to business needs quicker (the value added by a solution that provides agile capabilities applicable beyond the project under review)

    - Ability to roll out a solution faster (time-to-market benefits for IT such as what it takes to build a solution vs. an option that can be rolled out faster)

    - Patching, provisioning, issue resolution improvements (effective maintenance of the environment)

    - Enhanced monitoring and diagnosis capabilities (built-in and available tools in the ecosystem for delivering superior monitoring and integration capabilities)

    - Ability to integrate into the organizational ecosystem at the enterprise level vs. limited project level integration (such as a solution that readily integrates with the enterprise level information architecture)

    - Enhanced service level agreement (SLA) metrics

- Benefits from additional architectural factors such as reliability, scalability, and standardization (Since these factors are heavily weighted by a defined technical solution, they should be revisited often as specifications change. Sometimes, options being considered can have transformative capabilities in these areas and add significant value to the organization.)

- IT Employee Productivity Improvements:

  - Administrative effort reduction due to IT process improvements (Less time required to perform initial installation, testing, and integration to ongoing administration and maintenance, freeing up time for higher value activities)

  - Ability to scale platforms with fixed resources (vs. a need to add resources as platforms scale)

- Cost Avoidance:

  - Lower CapEx and OpEx for an option vs. others under consideration

  - Future compute and storage purchase avoidance due to the transformational nature of the solutions

  - SLA or other penalty avoidance

This list is only representative in nature. The IT benefits vary significantly from organization to organization. Table 3-3 illustrates a simple example of IT benefits value quantification.

*Table 3-3.* *Sample Summary IT Value Portion of a Business Case*

| Benefits | Total Cost | Savings / Year |
| --- | --- | --- |
| Reduced time to provision, monitor, tune, and manage the environment | Staff hours spent provisioning, monitoring, tuning: 2000*$65 = $130,000 (One person dedicated annually) | 10% of Cost $13,000 |
| Pre-tested patch bundles reduce time to upgrade and patch, and provide less complexity and less manual integration | Staff hours spent in patching, integration, and correcting errors that occur during integration: 1000*$65 = $65,000 (Half FTE annually) | 20% of Cost $13,000 |

*(continued)*

70

***Table 3-3.*** (*continued*)

| Benefits | Total Cost | Savings / Year |
|---|---|---|
| Faster data access | Staff hours spent accessing data, running queries, and so forth: 4400*$65 = $286,000 (Assuming 20 employees @1 hour a day for 220 days a year. In case of better integrated solutions, higher benefits can be realized.) | 10% of Cost $28,600 |
| Reduced downtime and lost IT productivity | Number of occurrences in a year: 4*$42,530 = $170,120 (IT Productivity: Average cost of data center outages computed from past history or industry benchmarks) | 50% of Cost $85,060 |
| Interface improvements | Staff hours spent running queries overseas and in US: 1100*$65 = $71,500 (Five people spending one hour a day for 220 days a year overseas) | 100% of Cost $71,500 |
| Efficient and easier storage management (Provisioning, tuning, replication, cloning, performance monitoring) | Staff hours spent managing the environment: 5000* $65 = $325,000 (Benefits from superior storage strategy coupled with data and information strategy) | 40% of Cost $130,000 |
| Total | | $341,160 |

The IT benefits might not appear to be very substantial. This is not unusual for computations of this type. Transformative information architecture projects, such as those that include Big Data and IoT, have tremendous upside because of the business value they can provide (and you will see this later in the chapter). Cost savings in IT does not usually drive these kinds of projects.

Remember that quantification of IT and business benefits involves about 80% factual data and 20% guesswork due to the potential variations and softer benefits that might be considered. However, at this point, we are simply trying to understand the potential benefits that might be part of our business case. So, it is a good idea to be flexible about what is included at this stage.

---

■ **Tip**   As organizations embark on developing complex business cases, we recommend that at least three data points be gathered for each benefit claimed to ensure that benefits are reasonable and represent a possible range of values. The three points typically represent conservative, pragmatic, and aggressive benefit projections.

---

In many scenarios, expected benefits could vary by quite a bit depending on assumptions made. Table 3-4 shows how the same savings might be calculated across a range of scenarios using a three-point approach that defines conservative, pragmatic, and aggressive benefit projections:

***Table 3-4.***  *Sample Detailed Calculations for an IT Value Driver*

| Description | Conservative | Pragmatic | Aggressive |
|---|---|---|---|
| [a] Number of administrators that support the environment (Source: Model Input) | 1 | 1 | 1 |
| [b] Fully burdened labor cost for each administrator (Source: Model Input) | 130,000 | $130,000 | $130,000 |
| [c] Total labor cost for integration activity (Source: Calculation: [a]/[b]) | 130,000 | $130,000 | $130,000 |
| [d] Estimated % reduction due to higher set of benefits. Purpose-built option is pre-configured, integrated, tested, and certified solution with out-of-the-box performance characteristics. This minimizes various IT staff activities including provisioning, tuning, and diagnostics, while also allowing administrators to leverage existing knowledge. Additionally, the solution offers a single point of support, which can further reduce the amount of time spent troubleshooting issues. (Source: Model Input based on known customer information) | 5% | 10% | 20% |
| [e] Potential new labor costs for the activity (Source: Calculation: [c]*(1-[d])) | $123,500 | $117,000 | $104,000 |
| [f] Summary benefits (Source: Calculation: [c]-[e]) | $6,500 | $13,000 | $26,000 |

By considering three data points for all the projected IT and business benefits, organizations can develop integrated projected benefits with a reasonable spread for review, as shown in Figure 3-16.

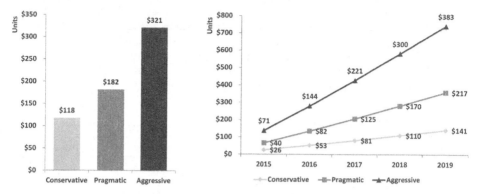

***Figure 3-16.*** *Sample projected benefits under three scenarios—conservative, pragmatic, and aggressive*

## Business Value

The process to identify and quantify the business value is similar to the process described for IT value quantification in the previous section. However, the categories and impact / value are significantly different and a lot more dependent on industry business drivers and corporate business initiatives. Business value tends to be much higher than IT value and is the main driver in the decision-making process.

Typically most of the business benefits can be again categorized into a handful of buckets, as discussed earlier in this chapter. Here is a list of some representative benefits that can be quantified as part of the business case development process:

- Ability to launch new business services

- Impact on revenue

- Cost avoidance and reductions

- Reduced business risk

- Business impact of unwanted planned and unplanned downtime

- Improved agility and responsiveness to customer and market demands

- Additional synergies across the organization due to strategic partnership between IT and organizational business

We will next take a look at a couple of examples based on common Big Data and Internet of Things initiatives we have observed in a couple of industries. The industries we will provide examples for are retail and manufacturing.

Almost every large retailer today is investing in an omni-channel strategy. This approach brings together management and analysis of brick-and-mortar stores and online stores in a continuum to create a seamless shopping experience. Because of the presence of web logs and the importance of shopper sentiment expressed through social media, these solutions often are designed to include Hadoop as part of the information architecture. Of course, the transactions in the various stores are tracked in relational databases, and data warehouses are part of the footprint. The future state information architecture can positively impact online revenue and the brick-and-mortar business due to the efficiency gains in the supply chain and operations. Table 3-5 provides a sample calculation describing the potential business impact to online sales:

*Table 3-5.* *Sample Detailed Calculations for the Online Sales Impact of a Retailer*

| Description | Conservative | Pragmatic | Aggressive |
|---|---|---|---|
| [a] Annual sales for the division (Source: Model Input. Recent annual report.) | $2,970,000,000 | $3,300,000,000 | $3,630,000,000 |
| [b] Percentage of sales through store brands (Source: Model Input. Per notes from workshops) | 36% | 40% | 44% |
| [c] Sales through brands in stores (Source: Calculation: [a]*[b]) | $1,069,200,000 | $1,320,000,000 | $1,597,200,000 |
| [d] Percentage of web sales off of items sold in stores (Source: Model Input. Per notes from workshops) | 18% | 20% | 22% |
| [e] Online sales revenue for the division (Source: Calculation: [c]*[d]. Also based on recent Q1 and Q2 results and previous annual report. This division contributes 71% to 78% of online commerce. This also aligns to the projected sales revenue.) | $192,456,000 | $264,000,000 | $351,384,000 |
| [f] Expected growth due to technology driven business impact (Source: Model Input) | 11% | 12% | 13% |
| [g] Digital sales impact projected due to technology driven innovation (Source: Calculation: [e]*[f]) | $20,785,248 | $31,680,000 | $46,382,688 |

Similarly, other business benefits can be quantified on an annual basis and then projected over a period of three to five years to show the true impact of the future state architecture.

In our second example, a manufacturer of industrial restaurant supplies is evaluating the value of analyzing data from sensors in freezers. The lines of business at the manufacturer see that they might use this technology to improve current sales of freezers and build repair business revenues. It is also believed that the life of the equipment could be extended in the process, improving customer satisfaction. Table 3-6 outlines the business impact.

*Table 3-6.* *Sample Detailed Calculations for the Business Impact through Freezer Servicing*

| Description | Conservative | Pragmatic | Aggressive |
| --- | --- | --- | --- |
| [a] Average annual market for freezers (Source: Model Input) | $40,500,000 | $45,000,000 | $49,500,000 |
| [b] Average cost of each freezer (Source: Model Input) | $4,500 | $5,000 | $5,500 |
| [c] Number of new freezers per year in the market (Source: Calculation [a]/[b]) | 9,000 | 9,000 | 9,000 |
| [d] Number of freezers sold by the organization in a year (Source: Model Input) | 900 | 1,000 | 1,100 |
| [e] Percentage of freezer market share (Source: Calculation: [d]/[c]) | 10% | 11% | 12% |
| [f] Expected percentage of freezer market share due to sensors (Source: Model Input) | 13% | 15% | 17% |
| [g] Additional freezers sold in a year due to impact of sensors (Source: Calculation: ([c]*[f])-[d]) | 248 | 350 | 385 |
| [h] Additional freezer revenue due to sensor impact (Source: Calculation: [b]/[g]) | $4,050,000 | $1,750,000 | $2,117,500 |
| [i] Average routine service visits needed per freezer (Source: Model Input) | 3 | 4 | 5 |
| [j] Charge for each routine service visit (Source: Model Input) | $45 | $50 | $55 |
| [k] Projected number of parts sold during visits (Source: Calculation: [g]/[i]) | 744 | 1,400 | 1,925 |
| [l] Potential additional parts revenue through new freezers (Source: Calculation: [j]/[k]) | $33,480 | $70,000 | $105,875 |
| [m] Average annual market for freezer service (Source: Model Input) | $13,500,000 | $15,000,000 | $16,500,000 |
| [n] Average service revenue per freezer (Source: Calculation: [m]/[c]) | $1,500 | $1,667 | $1,834 |
| [o] Additional service revenue through new freezer (Source: Calculation: [n]*[g]) | $372,000 | $583,450 | $706,090 |
| [p] Annual Summary Benefits (Source: Calculation: [h]+[l]+[o]) | $4,455,480 | $2,403,450 | $2,929,465 |

Quantification of anticipated benefits for initiatives under consideration enables organizations to compare costs against benefits over a specific time period and make informed investment decisions.

While the three-point approach for computing a range of benefits is extremely useful, for some scenarios organizations need much more granularity and need to calculate the upper and lower boundaries with specific probabilities. Monte Carlo methods can be used to run hundreds to thousands of such scenarios. Commercial software tools are often used to run these simulations, enabling an organization to determine risk and project benefits. This approach might be applied in cases where the order of magnitude for the investments is great or the number of inputs is very high with significant ranges of values. The cost of simulation software and the need for projecting the range of benefits at certain probability levels should be weighed against the simplicity and faster time to execute the three-point approach when deciding which approach is right for an organization.

Figure 3-17 is an example of output obtained from a software package that ran a Monte Carlo simulation on top of a business case model developed to support a potential initiative.

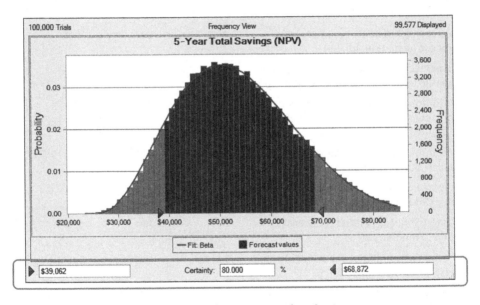

*Figure 3-17. Sample Monte Carlo simulations output for a business case*

As highlighted in Figure 3-17, an approach like this helps identify both the upper and lower boundaries of projected benefits with a specific certainty level. By adjusting this certainty level to acceptable probabilities, organizations can see the lower and upper boundaries of possible benefits before making an informed investment decision.

# Other Trade-offs to Consider

As we build the business case for these types of projects, we might begin to discover additional costs associated with creating enterprise class platforms for newer data management systems. We might also see a need for much faster time to market for the future state architecture because of the compelling nature of the business case.

A first challenge could be locating the skilled individuals needed to build and use the technology footprints under consideration. For example, the data scientists that are often associated with Hadoop remain scarce and expensive. Fortunately, there are many more ways of accessing and analyzing data in Hadoop today, including the use of SQL interfaces and popular predictive analytics tools. For ETL offloading, many of the common ETL tools now fully support Hadoop.

The need for faster enterprise class deployment and operational support might also cause us to consider evaluating purchase of pre-integrated server, storage, and software units as an alternative to Do-It-Yourself (DIY) configurations. Examples include an increasing array of appliances that are available for the various data management systems typical in the evolving information architecture.

Many Hadoop clusters start out with a relatively small number of nodes in labs or in similar small footprints in production environments. As the demand for larger clusters increases, operational efforts and associated risk go up as well. Organizations can be severely constrained for precious resources and reach that critical point quickly. Figure 3-18 illustrates the challenge.

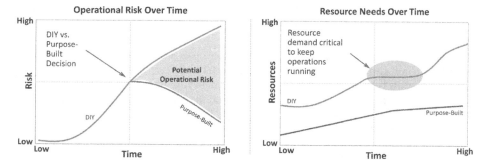

***Figure 3-18.*** *Do-It-Yourself resource considerations with growing Big Data environments*

Another option that might be considered is a cloud deployment strategy. Cloud-based solutions seem rather simple in comparison to assembling your own system, as illustrated in Figure 3-19. There can be significant cost and time-to-market benefits in taking a cloud-based approach.

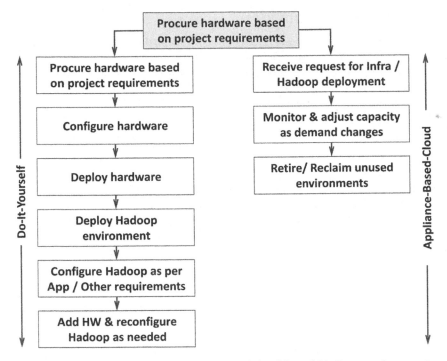

*Figure 3-19. Operational impact for DIY and cloud-based Big Data environments*

The ability to quickly scale up or down in response to business needs and the ability to provision quickly by using a self-service approach can be extremely desirable. Trade-offs are generally technical, especially where large data volumes might exist in locations widely separated from the cloud-based solution and there is a need to leverage all of the sources of data together to find business solutions. Then, discussions about network bandwidth and flexibility come first to mind and the costs associated with solving those issues become more apparent.

## We Have Only Just Begun

We have only just begun our business case. At this point, we now probably have a much stronger conviction that we can justify one or more of our projects. And those projects will drive the need for defining and deploying a future state architecture. However, we also realize that we will need to revisit the business case later in our methodology after we have gathered a lot more information.

Now that some of the business drivers are identified and we have aligned them to prioritized initiatives, we are next going to identify the potential data sources we need. We will also gain a better understanding of our reporting, query, and analytic needs and how the data will help answer some of the business challenges we identified in this chapter.

# CHAPTER 4

■ ■ ■

# Business Information Mapping for Big Data and Internet of Things

Thus far, we have defined a broad business and IT vision of a potential project as we traversed through our methodology for success. We should now also have a more detailed understanding as to the potential business goals for our project since we fused the vision with a more detailed strategy and better understanding of the organization's goals. Stakeholders and potential sponsors are now being identified. However, we are still quite far from understanding necessary data flow and processing changes that will impact our current information architecture.

In this chapter, we will map the key performance indicators (KPIs) that we need to run the business back to their data sources. We will also begin to understand the gaps we have in data, the analysis needed, and how our data might be represented to the business. Our business information maps (or BIMs as sometimes referred in this chapter) will help drive our technology footprint redesign in the next phase of our methodology for success. But here, you might ask why we develop a series of maps to do this.

Mapping of the data flows will help us in many ways. There is an old adage that "a picture is worth a thousand words." Graphical drawings can simplify the communication of complex ideas. This is certainly true when we use our finger to trace how an object will move between two points on a map. For millennia, we have used such maps to guide journeys.

In some ways, our use of maps to understand the complexities of the world is made possible by our brains' hardwiring for processing two-dimensional images gathered by our eyes. In fact, today's neuroscientists believe that over half of our brain is dedicated to processing vision. That is why we can so easily understand and spot patterns.

One could argue that our brains are already adapted to processing the visual output from Big Data analysis in this way. Such output is often presented in graphical form. For example, lift charts that compare outcomes from random selections vs. system selections help us better understand the effectiveness of prediction.

In comparison, it has only been relatively recent in human evolution that we've had language. Even more recently, we've adapted to understand written language through years of training in schools. Hence, you are able to understand the methodology and details that we are describing this book (we hope).

Multi-dimensional pictures can communicate far more information than uni-dimensional text. They are extremely useful in breaking down barriers that cause misunderstandings created by language and culture. This is especially true when we use commonly accepted and recognized notations and graphical icons. Consequently, in this chapter, we will call upon a recognized standard for mapping data flows: Yourdon / DeMarco. We will be drawing (both literally and figuratively) from this standard to provide visual analysis.

Since this book concerns itself with the architecture of Big Data and the Internet of Things, we will do what architects in the physical world would do. We will create images that can be understood by our intended audience and that communicate relevant information in a format that is easy to absorb.

Figure 4-1 illustrates our current phase in our methodology for success. It also outlines the key topics covered in this chapter.

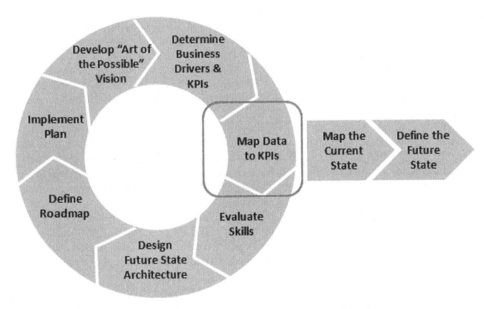

*Figure 4-1. Business information mapping phase in our methodology for success*

The chapter is divided into two sections. In the first section, our primary focus is on developing the current state business information map. The second section walks through the process of defining the future state and mapping it onto a BIM representing that future.

# Mapping the Current State

Before providing an example of how to go about creating a current state business information map, we begin by providing some background on the Yourdon / DeMarco notations commonly used to define the components in data flow diagrams that we use here. This will help you understand what the figures in this chapter represent and also establish standards for the maps you will later develop on your own.

We also will take another look at how our output from the previous phases of the methodology will help us build business information maps. We will review a theoretical business situation that would have been uncovered previously in our methodology and use the background knowledge that we have gained to develop a BIM representing the current state.

## Data Flow Diagram Basics

Data flow diagrams (DFDs) are appropriately named as they are used to graphically represent input into a system and the resulting output. They are also used to illustrate key processes and data stores. Edward Yourdon and Tom DeMarco created a set of standard objects that are the components used in typical data flow diagrams.

Using the Yourdon / DeMarco definitions, the four basic DFD components are described as follows:

- Input / Output: A person, organization, or system that is external to the system, but interacts with it.

- Process: An activity or a function that is performed for some specific reason. It can be manual or computerized. Each process should perform only one activity.

- Data Store: A collection of data that is permanently stored.

- Data Flow: A single piece of data or logical collection of information that is moving between any of the above.

Figure 4-2 illustrates the symbols or icons for these four components using Yourdon / DeMarco notation for DFDs.

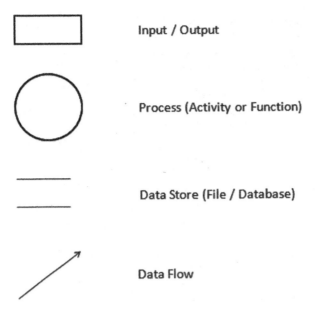

*Figure 4-2.* *Yourdon / DeMarco notation*

DFDs are used to visually represent where data comes from, how systems process data, and the output from processing. The systems might already exist (current state) or be in the planning stages (future state). Systems analysts typically create DFD diagrams by first meeting with business analysts and architects. After the diagrams are created, they ask the same individuals to validate that the drawings match reality or the desired outcome.

The following are considered as best practices for creating data flow diagrams:

- A series of data flows always start or end at an input / output and start or end at a data store. Conversely, this means that a series of data flows cannot start or end at a process.

- A process must have both data inflows and outflows.

- All data flows must be labeled with the precise data that is being exchanged.

- Process names should start with a verb and end with a noun.

- Data flows are named as descriptive nouns.

- A data store must have at least one data inflow.

- A data flow cannot go between an input / output and a data store. A process must be in between.

- A data flow cannot go between two data stores. A process must be in between.

- Inputs / outputs and data flows can be repeated on a data flow diagram in order to avoid lines crossing, but processes are not repeated.

The first step in developing a DFD is to create a context diagram and name it after the system being modeled. Sometimes, this diagram is called a Level Zero (0) diagram since there is no higher level of abstraction. It represents the entire system as a single process and data flows to inputs / outputs with which the system interacts. Unlike other lower-level processes that are defined (such as Level 1, 2, and so on), the context process is usually the name of a system. Lower-level processes usually have names starting with verb-like identifiers. Typically, the context diagram does not picture the data stores as they are internal to the system.

Figure 4-3 illustrates a typical example of a context diagram or Level 0 DFD.

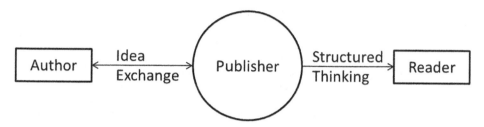

***Figure 4-3.*** *Example context (Level 0) DFD entitled "Publisher"*

You might think of the context process circle as a manhole that descends to an underlying system being analyzed. At this top level of abstraction, the data flows are like cables that disappear into and emerge from this manhole. To further understand the communication and processing that goes inside the context process, we need to descend at least one level down into our subterranean world to follow our data flow cables. The first level down is called Level 1. Here we can trace data cables that came in and out through the manhole at Level 0 as we see where they go through additional sub-processes. These sub-processes are themselves like manholes that may descend to additional levels of detail. Some of the data cables that come out of these subterranean manholes will go to data stores or reservoirs (files and databases). These reservoirs may be connected by one or more data cables to other processes.

For example, if we decomposed the context DFD in Figure 4-3 into a Level 1 DFD, it might look like the diagram illustrated in Figure 4-4.

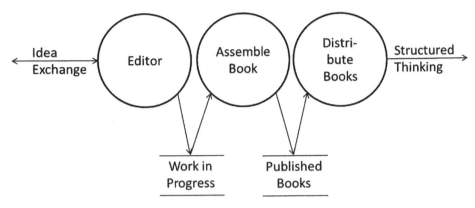

*Figure 4-4. Example (Level 1) DFD entitled "Publisher"*

Thus far, we have introduced generic Yourdon / DeMarco concepts. At this point, we introduce the key performance indicator (KPI) symbol as an extremely important notation in our business information map. Recall that KPIs are the critical metrics required by business analysts or business processes to make business decisions. The KPIs often appear in executive dashboards. In Figure 4-5, we illustrate how we will represent them in our BIM diagrams.

*Figure 4-5. Key performance indicator as represented in our BIM diagrams*

## Understanding the Current Situation

In Chapter 3, we described a process that included understanding the business needs, identifying critical success factors, and also understanding the measures and KPIs needed to provide a solution to the current business problem. We go through that process prior to creating our business information maps. To set the stage for their creation, we will illustrate gathering of this information from a fictitious manufacturer of luxury cars, Lux Motor Cars (LMC), as an example.

LMC markets fine automobiles all over the world. In the United States, 70% to 75% of the new cars it sells are financed using an in-house leasing company, LMC Lease. The percentage of cars being leased is growing. Customers who use this financing option love the following features:

- The low, predictable cost and not having to come up with any "upfront" money. In other words, the leasing company actually buys the car and effectively rents it to the leaseholder for a fixed price over a fixed period of time to smooth out both the vehicle depreciation and the effective interest on the capital outlay.

- The fact that the manufacturer assumes the risk of major repairs through a new car warranty.

To assure that vehicles were competitively priced in the market, LMC kept gross margins on the initial vehicle sales quite low. This strategy was a viable business approach because the vehicle quality was high and the resulting warranty costs were much lower than the industry average. Since the cars stayed on the road a long time, LMC ultimately made the bulk of its profit by maintaining higher margins on vehicle parts and proprietary lubricants consumed during scheduled maintenance. LMC knows that the best way to keep profits up and warranty costs down is to encourage the owners to be pro-active with maintenance (keeping fluid levels topped up, changing of oil filters at proper intervals, and so on) with LMC-branded products. Consequently, industry data shows that LMC has the lowest total cost of ownership in each class of vehicles that it offers.

While the leasing numbers are continuing to grow, LMC management is seeing a disturbing trend. Not only are LMC customer quality ratings declining in the United States, but the warranty costs are going up, too. Upon further analysis, the vehicles owned by LMC Lease in the United States are causing the unexpected increase in warranty costs while the lessees' satisfaction ratings are dropping.

Using information discovery tools to sift through the service records of all of the LMC Lease vehicles in the United States, the following information is uncovered:

- The variance between when a car is due for scheduled service and when it is actually serviced is quite high. In other words, lessees don't seem to adhere to the service schedule as closely as other LMC owners do.

- The number of miles driven after a "Check Engine" light came on was dramatically higher for LMC Lease vehicles. It appears that lessees weren't taking the "Check Engine" light very seriously.

The lessees seem to be much more likely to treat a LMC Lease car as if it were a rental car. Interviews held with lessees who had the highest claims uncovered the following about them:

- The lessees considered themselves to be very busy people and said they responded quickly to the "Service Due" light by calling the service department at their LMC dealer. However, when faced with the choice of coming in immediately for scheduled service vs. waiting for when a loaner car could be arranged, they chose to wait. Apparently, they valued the continuous availability of a vehicle.

- The most expensive warranty claims came from lessee vehicles that had a "Check Engine" light on and the lessee drivers said they felt the car wasn't working quite right. But again, when they called for service, they chose to wait until there was a loaner car available. They didn't want to rent a replacement vehicle, especially if they didn't know how long they would need it for.

Let's now review the information we have gathered so far. Two key LMC business goals stand out: decreasing the escalating costs of warranties in the United States and increasing the vehicle quality ratings among lessees.

Upon further discussion, we find that the warranty support organization believes that 30% to 50% of the cost of warranty claims could be avoided if LMC Lease cars had a telematics system that reported problems and automatically scheduled the vehicles for servicing when problems were reported. Such a system could enable the service group to better optimize staffing levels in the repair garages. The car fleet management group could better optimize the number of loaner cars and could augment the loaners with rental cars as appropriate. Finally, the company knows that customers would pay a premium for such a service from market research data that was gathered.

There are a number of KPIs that the lines of business clearly want to track. These KPIs include lifetime cost of warranty, lifetime vehicle profitability, service scheduling wait time, loaner car availability and utilization, service garage profitability, premium service profitability, and customer satisfaction.

LMC is currently not able to put such a program into place. Much of the data they need to run a program of this type is not currently available, though this is not fully understood in the business. So, we will next build business information maps describing the current state so that all can better understand how this part of the business operates with the data that it has today.

The good news is that LMC is ready to make a telematics investment. We will explore the potential impact of that investment later in this chapter. There, we will look at how business processes can be changed when that data becomes available in the future state.

# Building a Current State Business Information Map

The BIMs are typically developed through a collaboration of knowledgeable business executives and analysts, data scientists, and IT (particularly if the business analysts and data scientists are unsure about where the data might be sourced from). The lines of business executives provide important information as to how they make decisions using the data provided.

---

■ **Note** The IT organization will likely want to provide their view of data flows in the current state. That is fine. However, in many organizations, the lines of business will have created additional reports incorporating undocumented data sources. So we should interview all of the interested parties to truly understand the current state of data and how it is used.

---

You might wonder why we will build the BIMs for the current state prior to laying out a future state diagram. The current state provides us with a baseline understanding of the data sources that are currently available and how they are being utilized. This understanding can include the types of data stores, data integration methods in the current footprint, and the reporting and analytics capabilities. We will build upon this baseline as we define the future state.

In our LMC example, we will want to talk to key executives and business analysts from multiple areas of the business. Conducting this discovery through a workshop is a common approach. Information must be gathered from product development, leasing, warranty support, dealer servicing, and the car fleet management lines of business. We need to understand how the current data feeds are used when making decisions, what intelligence they provide, and current limitations.

To assure a strong opening in our first BIM exercise, it is a good idea to have the potential business sponsor kick off the meeting. For a company like LMC, we might want to have the VP of Vehicle Product Development or a similarly titled person open the discussion. This helps drive home the importance of the effort.

During the workshop or interviews, we will first validate how various parts of the organization view the opportunity and current situation. For LMC, the following provides an example of what we might discover:

- From our leasing team interviews, we discover that while we understand the credit worthiness of our customers, we also have found that many lessees see the lease as the same as "automobile as a service." A common request is for an extended service warranty at a premium price that guarantees availability of loaner vehicles when needed.

- Our warranty support team believes that sensor logs in the vehicles hold the key to managing costs. Today, the on-board systems in the cars alert drivers to possible problems and what to do (for example, service due, check engine, and so on). The sensors produce error codes that are stored in the vehicle, and this data is downloaded when the vehicle goes in for service. By processing the data in real time, it is believed that major vehicle failures could be avoided and cost of service could be reduced.

- The service organization sees the lack of loaner cars as delaying timely service. Their only contact with the customer is when the car is brought in. They believe that more timely service could also reduce warranty costs.

- The car fleet management organization only measures the percent of loaner cars that are utilized. Since the percentage is high (for example, 95%), they feel as though they were providing a well-optimized service.

- Executives indicate that a lack of visible KPIs for warranty support is especially troubling.

Now let's take a look at how the LMC maintenance and warranty (M&W) system currently operates. We start by representing the current state business information map at its highest level (Level 0). First, we identify the potential data sources:

- Sensors: The sensors are in LMC automobiles owned by LMC Lease. The M&W system gathers data from the sensors, but doesn't provide feedback.

- Garage: The garage represents the LMC dealership service department. The M&W system interacts with the various service departments, but how they operate is beyond the scope of this analysis.

- Loaner: This represents our car fleet management loaner data. Although the lack of availability of loaner vehicles is an issue, how this department operates is also beyond the scope of this analysis. For completeness, we will show the relationship to the M&W system.

- Lessee: This represents the driver / lessee of an LMC Lease vehicle. LMC Lease's customers driving the vehicles are clearly at the heart of LMC's business, but only at the edge of this analysis. Once again, for completeness, we will show the relationship to the M&W system.

- Stakeholders: This represents LMC executives who care about the performance of the M&W system. We will indicate that there is not much in the way of KPIs flowing back to those stakeholders now.

These sources of data are represented in Figure 4-6. We are not illustrating data stores that are internal to the M&W system (such as a vehicle database or data from on-board electronics) at this point.

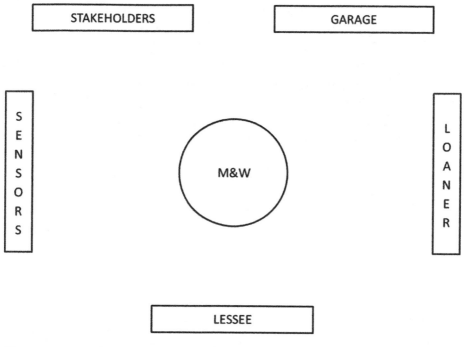

***Figure 4-6.*** *Partial current state context (Level 0) BIM entitled "LMC Maintenance and Warranty System"*

Let's next evaluate how the data flows from these sources. We represent the following data flows in Figure 4-7:

- Sensors send batch diagnostic logs through the automotive diagnostic scanner port under the dashboard (batch diagnostics).

- Sensors send real-time data that is processed in the on-board electronics portion of the M&W system (R T sensor measurements).

- Lessee and / or driver receive information from the on-board electronics portion of the M&W system in the form of real-time idiot lights and gauges (R T driver indicators).

- Lessee and / or driver exchange information with the M&W system by doing automotive service scheduling negotiation (lessee scheduling).

- Lessee and / or driver receive the non-warranty portion of a service invoice (lessee invoice).

- Loaner car fleet management exchanges information with the M&W system in executing loaner car scheduling negotiation (loaner scheduling).

- Loaner car fleet management sends to the M&W system the loaner invoices for cost accounting (loaner invoice).

- Garage receives from the M&W system the vehicle data including diagnostics and service recommendations (servicing data).

- Garage exchanges information with the M&W system and performs servicing scheduling negotiation (Garage Scheduling).

- Garage sends the M&W system the full service invoice including the warranty and non-warranty portions (total invoice).

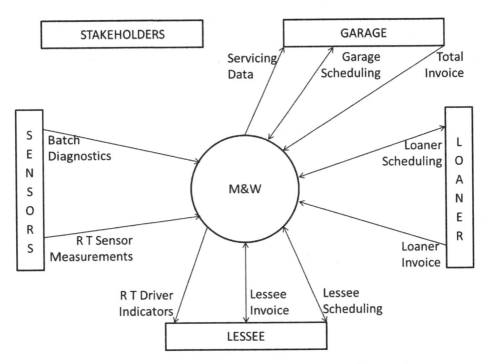

*Figure 4-7.* *Completed current state context (Level 0) BIM entitled "LMC Maintenance and Warranty System"*

As you can see, we illustrated all of these data flows in a single fairly simple diagram. The business information map begins to tell a story of how our current maintenance and warranty system operates. Now, we are ready to explore how the system operates at a level deeper.

We begin with the realization that the on-board electronics part of the M&W (that interprets the real-time (RT) sensor measurements to create the real-time (RT) driver indicators is completely disconnected from the rest of the M&W system. So, the first process shown on a Level 1 diagram is named "Interpret Sensors," as illustrated in Figure 4-8.

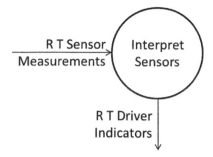

***Figure 4-8.*** *Partial current state (Level 1) "LMC M&W System" BIM (on-car processing only)*

Another singular process, "Empower Garage," enables the service department to repair LMC vehicles by converting batch diagnostics into actionable servicing Data. It is linked to the vehicle database. We illustrate this addition in Figure 4-9.

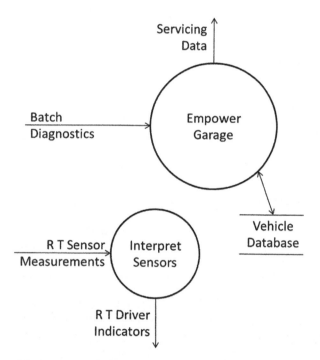

***Figure 4-9.*** *Partial current state (Level 1) LMC M&W system BIM (on-car processing and car/gargage interaction)*

The remainder of the current state M&W system negotiates scheduling and handles the accounting for vehicle servicing. So we show the "Account and Schedule Service" process and add a link from it to the vehicle database in our diagram. The completed Level 1 BIM is illustrated in Figure 4-10.

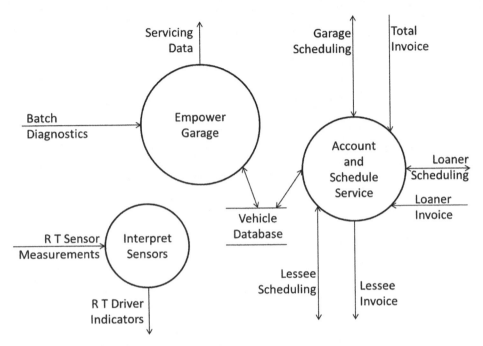

***Figure 4-10.*** *Completed current state (Level 1) "LMC M&W System" BIM*

# Defining the Future State

From our current state business information map, we can clearly see that we can't get access to the KPIs we identified earlier as necessary to optimally run the car maintenance and warranty business. As we define the future state with the help of the team that we gathered, we will determine how the data must flow in order to provide the business with the critical missing pieces of information.

In this section of the chapter, we will briefly cover preparing for the future state BIM exercise, continuing to discuss our LMC example as we define the future state, and then begin the transition as to how we might fit technology platforms onto our business information map.

## Preparing for a Future State Meeting

As we reconvene the team, we will first review our current state BIMs. We do this to make sure we didn't miss anything, but also to remind participants of what was discussed previously. As before, we will invite the appropriate business executives and analysts, data scientists, and IT data managers. As sensors will play a critical role in how we will gather data in our future state and additional functionality is likely to be envisioned, we will want to add engineering representatives to our discussion of the future state BIM.

---

■ **Note**   The notion that a project might possibly become funded can lead to a gain in momentum and growing general interest during this phase. Sometimes, a much more diverse audience becomes interested in the project and wants to join the discussion. To keep the BIM exercise manageable, separate briefings are sometimes held with newer members of the team to review the information that was previously gathered and to avoid needless review for those who helped formulate the earlier requirements and current state BIM.

---

Recall that earlier in this chapter, we mentioned the KPIs that would become significant to running the business in its future state. These included lifetime cost of warranty, lifetime vehicle profitability, service scheduling wait time, loaner car availability and utilization, service garage profitability, premium service profitability, and customer satisfaction. We also saw the need to provide a more automated maintenance scheduling system. So now, we will start to identify the data that we need and the flow of the data and processes that will provide the needed information and enable the capabilities that the business requires.

## The Future State Business Information Map

In the current state BIM diagrams, you probably noticed a dearth of KPIs indicated. Clearly, LMC doesn't have access to these KPIs needed to change how they run the business. Other key linkages and processes for providing automated maintenance scheduling were also missing.

In developing our future state BIM, we begin by looking at KPIs importance to stakeholders and how we will add new data flows for sensors and lessees. Additional data flows will be required as follows:

- Stakeholders will receive from the M&W system the key performance indicators (KPIs) that have been heretofore unavailable.

- Sensors will exchange data with the M&W system including the interactive diagnostics that include both driver alerts and on-demand logs (interactive diagnostics).

- Lessee and / or drivers will receive near real-time (RT) driver alerts and exchange messages to deal with service scheduling (near RT driver alerts and scheduling).

A future state context (Level 0) BIM illustrating these additional data flows, indicated by the shading, could be drawn as illustrated in Figure 4-11.

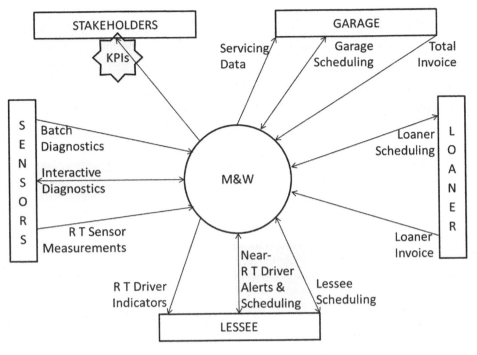

***Figure 4-11.*** *Future state context (Level 0) BIM for LMC M&W system*

Next, we take another look at the current state (Level 1) BIM produced earlier and make a few modifications. We rename the "Empower Garage" process to "Optimize Maintenance," based on our project's proposed desired outcome. We also add interactions with service scheduling and the LMC accounting system as follows:

- Optimized scheduling to negotiate scheduling the garage and loaner car at an optimal time.

- Accounting KPIs flowing back into the optimize-maintenance process to make sure the choices being made optimize composite KPIs of both processes.

Consequently, the future state context (Level 1) BIM could be represented as shown in Figure 4-12, with the shading indicating how the additional data flows are being processed.

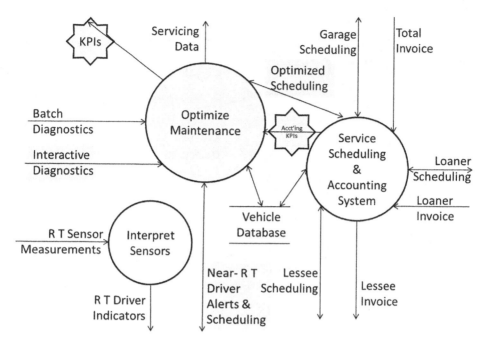

**Figure 4-12.** *Future state (Level ) 1 "LMC M&W System" BIM*

Since the optimize-maintenance process is so important to our project's success, we will next drill into how it will work. We will create a Level 2 diagram.

In order to begin automation of this process, sensors in the next generation of LMC vehicles will have the capability of streaming alerts through a telematics system to a vehicle event database optimized for event processing. A "trouble" event is detected by an optimize–short-term–event-maintenance process and more information will be requested interactively from on-board sensors in order to create a proposed action plan.

Data is forwarded as a maintenance event detected to a long-term–event-maintenance process that evaluates the event in the context of similar historical events and the most economical way of solving the problem. This information drives automatic negotiating of when the vehicle can be seen by service to resolve the issue and a reservation for a loaner vehicle consistent with LMC's business goals. Though an indicator light might come on in real time, the negotiation occurs in the background in near real time.

In order for the system to become smarter over time, event and other relevant data will be archived for later use into the vehicle log database. To reduce telematics costs, normal logs not associated with alerts will be downloaded the next time the car is plugged into the automotive scanner when visiting service and added to the same vehicle log database. LMC will get visibility into how the optimize-maintenance process is working through reporting KPIs.

Figure 4-13 illustrates the Level 2 Optimize Maintenance BIM for the scenarios that we just described, again with the shading indicating how the additional data flows are being processed.

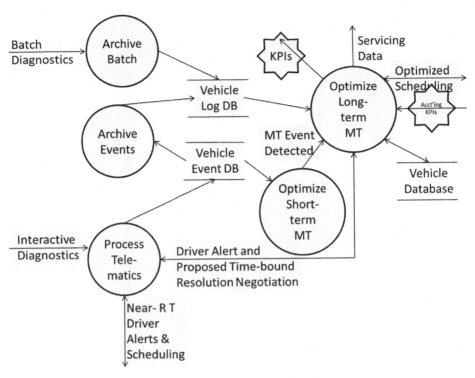

***Figure 4-13.*** *Future state (Level 2) "Optimize Maintenance" BIM*

# Transitioning to the Technology Design

As we did previously with our current state BIMs, we will validate our future state diagrams in follow-up sessions with a variety of audiences present. We will look for gaps between needed KPIs and the KPIs that the diagrams indicate are possible. We also look for gaps caused by any lack of additional data sources and processes that key stakeholders insist are needed. If we find such gaps, we will adjust our diagrams and gain consensus.

The business information maps will have an important role in helping us define the technology behind our information architecture in Chapter 6. As we make the transition to this next phase, we begin sharing the diagrams with architecture design teams. They will likely have views about the data management systems and tools needed to deliver the solution. Your BIM diagram might begin to resemble Figure 4-14 when the teams start to evaluate the impact of the data flows and the processing needed. The technology platforms are called out in the diagonal labels that are pictured.

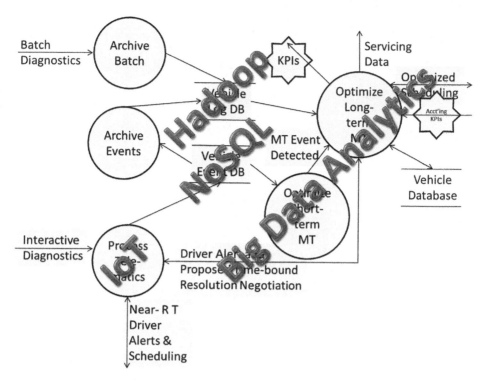

***Figure 4-14.*** *Future state (Level 2) "Optimize Maintenance" BIM with overlayed Big Data and IoT technology*

When we have consensus, we are almost ready to begin our more detailed technical architecture design work. A goal during that phase will be to begin to understand how big, complex, and expensive the project might actually become. Much of this design work will be driven by the need to eliminate the data and processing gaps in our current information architecture that we exposed when we created the future state BIMs in this phase.

However, before we begin the technical architecture design work, we will find it useful to understand the skills that we have currently available and identify critical skills that are lacking. This can help us better determine how to best solve the data and processing gaps that we just uncovered.

# CHAPTER 5

■ ■ ■

# Understanding Organizational Skills

In this chapter, we explore assessing the skills present in an organization that are critical to the successful design, deployment, and management of the future state information architecture for Big Data and the Internet of Things projects. Covering the topic of skills at this point in the book might cause you to scratch your head. Isn't it premature to be evaluating the skills in our organization before we define in some detail the future state information architecture? After all, the information architecture should match the business requirements, and we described how to gather those requirements in the previous chapters. We also described creating a broad vision of what the future architecture could look like much earlier in our methodology. We might assume that we can now take the business requirements and begin planning details about the future state information architecture.

However, if we jump into planning the future state now, we could find that we are missing some critical information that we should consider. More often than not, more than one information architecture technology footprint might provide a capable solution for the same business problem. Some of the footprints we will consider will certainly be more technically elegant than others. But we might want to consider whether suggesting a footprint that we would have difficulty implementing or maintaining is wise. If we have a choice of information architecture designs and the technical trade-offs are not too dire, we might want to reconsider our choice if we believe that meeting the skills requirements will be difficult and unique to a specific footprint and these skills will not be needed as part of a wider strategic platform vision. Of course, we won't be able to fully evaluate architecture trade-offs until we begin to understand the skills that we do have present in our organization.

That said, after we define our future state information architecture in more detail, we will have a more complete view as to the skills that are required (and those that are not). We will begin to understand the vendor and open source products that might be implemented, so we will need to explore the skills required in more detail as well to better understand our true implementation costs. You will see that we revisit the skills discussion in Chapter 7 of this book with a more detailed focus on the Hadoop, NoSQL database, sensor development, and network communications backbone skills required.

As we begin the skills evaluation process here, we need to have a framework that describes the key skills we are evaluating and are assigning metrics to. In this chapter, we will describe how we can evaluate the skills in alignment with the four types of architecture that are described by TOGAF (as introduced in Chapter 1). Those four areas are business architecture, data architecture, application architecture, and technology architecture. We will introduce many of the critical skills needed in the organization for successful design, deployment, and ongoing utilization and management for each architecture type.

Figure 5-1 highlights the phase we are at in our methodology for project success and what we are covering in this chapter. As we evaluate the skills in an organization, we will look at both the business skills and the technical skills required. And we will begin to understand critical gaps that the organization currently has in skills that could impact our ability to successfully deploy, manage, and utilize our future state architecture.

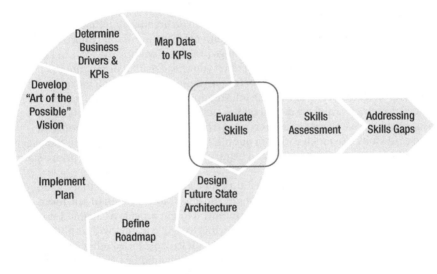

***Figure 5-1.*** *Skills evaluation phase in our methodology for success*

Next, we will take a look at the metrics we might use when assessing skills. We then provide summarized descriptions of many of the key skills that could be required. Finally, we cover delivering the news regarding skills gaps, including validating what we uncovered during our assessment, and some of the choices we will have in remediating the gaps.

# Skills Assessment and Metrics

The skills assessment usually consists of a series of interviews in the lines of business and departments that are critical to our project succeeding. The interviews are generally held with managers of the departments who are knowledgeable about the skills of their employees. In addition to identifying whether anyone in the organization has a certain skill, we also want to understand how widespread the skills are during this process. After all, if we have a certain critical skill present in only one individual, this is going to severely limit our ability to scale the effort on the project or scale the number of projects that we can tackle.

In order to assess and quantify the maturity of the skills that are present, we suggest using a simple scale of zero to five. More often than not, this exercise is facilitated using spreadsheets where the metrics are gathered that we describe. Some also use the spreadsheet to record a desired future skill level of maturity in addition to the current rating. The spreadsheet can then be used to show where skills gaps exist that could get in the way of project success and how wide they are.

For the evaluation of skills maturity, Table 5-1 contains definitions that can be used to assess the skills present in the organization. Values for the ratings numbers represent a range from no skills present to very advanced skills widely available, including for transformational activities. You might choose to modify the definitions and descriptions of the skills metrics presented here based on your own unique needs, but this table at least provides a starting point.

***Table 5-1.*** *Skills Assessment Metrics*

| Rating | Definition | Where Skill Applied |
|---|---|---|
| 0 | No skills found anywhere within organization | Skill has not been applied or, if applied, was outsourced |
| 1 | Skill is in a very early stage of development, planned or very limited | Skill development typically through prototype building |
| 2 | Skill has been developed on a very tactical level | Skill is directly tied to a project |
| 3 | Skill is available and strategic for a localized set of current projects | Skill has been or will be applied multiple times at a department level |
| 4 | Skill available and strategic for a wide breadth of current projects | Skill has been or will be applied multiple times in the enterprise |
| 5 | Skill available and strategic for a wide breadth of transformational projects | Critical mass of skills available to meet innovative project needs across the enterprise |

Figure 5-2 illustrates a portion of a typical skills assessment spreadsheet. The skills assessment metrics and further explanations as to the meaning of the metrics for a technology architecture skills evaluation appear in the upper portion of our pictured spreadsheet. You can see that we provided a typical question that might be asked to determine a skill level and a place to enter the rating for the current state of each skill. We also have a place to enter the level of skill desired in our future state.

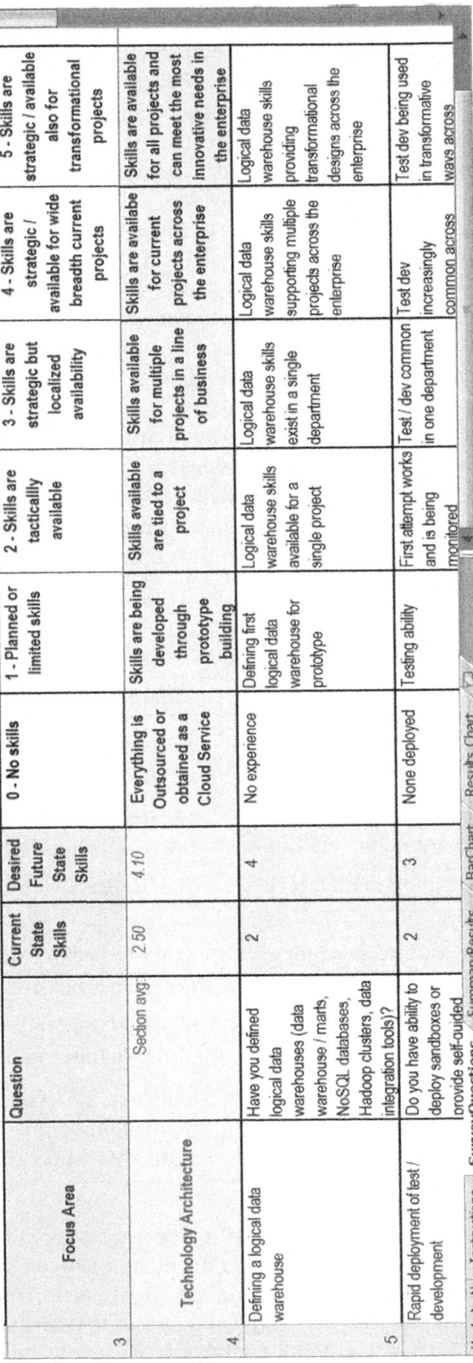

| Focus Area | Question | Current State Skills | Desired Future State Skills | 0 - No skills | 1 - Planned or limited skills | 2 - Skills are tacticallly available | 3 - Skills are strategic but localized availability | 4 - Skills are strategic / available for wide breadth current projects | 5 - Skills are strategic / available also for transformational projects |
|---|---|---|---|---|---|---|---|---|---|
| | Section avg | 2.50 | 4.10 | Everything is Outsourced or obtained as a Cloud Service | Skills are being developed through prototype building | Skills available are tied to a project | Skills available for multiple projects in a line of business | Skills are availabe for current projects across the enterprise | Skills are available for all projects and can meet the most innovative needs in the enterprise |
| Technology Architecture | | | | | | | | | |
| Defining a logical data warehouse | Have you defined logical data warehouses (data warehouse / marts, NoSQL databases, Hadoop clusters, data integration tools)? | 2 | 4 | No experience | Defining first logical data warehouse for prototype | Logical data warehouse skills available for a single project | Logical data warehouse skills exist in a single department | Logical data warehouse skills supporting multiple projects across the enterprise | Logical data warehouse skills providing transformational designs across the enterprise |
| Rapid deployment of test / development | Do you have ability to deploy sandboxes or provide self-guided | 2 | 3 | None deployed | Testing ability | First attempt works and is being monitored | Test / dev common in one department | Test dev increasingly common across | Test dev being used in transformative ways across |

Instructions  |  **SurveyQuestions**  |  SummaryResults  |  BarChart  |  Results Chart

*Figure 5-2. Skills assessment metrics in our spreadsheet*

Now that we have outlined how we will evaluate the skills, we will next take a look at the skills we might assess in alignment to the business architecture, data architecture, application architecture, and technology architecture for our proposed project.

# Business Architecture Skills

The set of skills we describe in this section is critical to successfully linking our future information architecture to the changing needs of the business. The business-related skills we evaluate within our organization typically include an ability to formulate business strategy and define key business processes, an organizational maturity in determining business requirements, and an ability to understand business mandates and objectives that drive data availability and governance needs.

For business strategy maturity, some of the skills we might evaluate in our organization include the following:

- Business plan development skills including the linkage of line of business objectives and goals to enterprise objectives and goals, as well as the prioritizing of the objectives based on business implications for the entire organization.

- The ability to understand critical success factors for business objectives and goals from various points of view within our organization.

- Experience in recognizing and uncovering the value and other benefits that will occur in reaching desired business objectives and an ability to build a business case that reflects that value. (Some of the underlying indicators of skills include having a defined business case methodology, an ability to perform net present value [NPV] calculations to compare options, and other accepted standards in the organization.)

- Maturity in establishing clear sponsorship, funding, and accountability measures for success where lines of business take part in defining information technology initiatives.

As we evaluate our ability to define and execute business processes, some of the skills we might evaluate in the organization include the following:

- Experience in the translation of business needs and goals into repeatable processes.

- A strong commitment to and experience in managing the business by using key performance indicators (KPIs) delivered in reports and an ability to define the KPIs and key measures that are required.

- Experience in managing the business through ad hoc query tools and the ability to articulate requirements for fact and dimensional data.

- The ability to use statistical analysis and data mining tools to understand past business performance and / or predict future outcomes.

- Experience in using information discovery tools to uncover the characteristics of data and identify new sources for reporting, ad hoc query analysis, and statistical analysis.

- The ability to use advanced programming tools and utilities (such as those often used by data scientists and commonly found in Hadoop environments including Java, Python, Ruby, MapReduce, and Spark).

---

■ **Note**    In Chapter 1, we suggested the typical relative user community sizes for information discovery, business intelligence, predictive analytics, and other tools and techniques used by business analysts and data scientists. When we perform a skills assessment, we can understand the true user community sizes in our organization.

---

When understanding our organization's maturity in gathering and understanding business requirements, some of the skills evaluated can be tied to structures and methodologies we have put in place. These can include the following:

- Experience in formally gathering requirements through competency centers linking lines of business needs to IT architects, planners, and developers.

- Maturity in managing business change when new technology is deployed, including our methods for education on the usage of the technology, in order to assure adoption.

We will also likely consider the ability to understand business needs and mandates that impact our data availability and governance approach. Some of the skills we can evaluate in our organization include the following:

- The ability to accurately assess and articulate what data must be available for analysis (for example, data description, granularity, and length of history) and to describe the allowable latency for data availability that ensures that timely decisions or recommendations can be made.

- Experience in assessing data quality and its impact on making accurate business decisions.

- Maturity in using data lineage to evaluate how data transformations and the data sources chosen might impact the data we are using to make business decisions.

- The ability to align the need for access to data with security standards and policies in deciding who should have data access and who should be denied access.

# Data Architecture Skills

The data architecture skills that are assessed usually include those required when defining and managing the logical and physical data structures that can exist within data management systems. The skills evaluated commonly include those needed to manage the data management systems themselves. As we discussed in earlier chapters, some of the key data management platforms deployed in Big Data and Internet of Things projects include relational databases, NoSQL databases, and Hadoop.

When assessing the skills required for defining and managing the data structures, some that we might focus on include the following:

- Maturity in choosing and defining the right data management systems that align to the data processing workloads that deliver the information that the business needs for solving their business problems (such as relational databases for structured data, NoSQL databases for ingestion scalability and lightweight transactions, Hadoop for processing varied data types and predictive analytics, and so on).

- Experience in defining relational database data models that align to business requirements and also provide agility (such as the use of third normal form where needed for EDWs, conformed dimensions across star schema models where those are deployed, and so on).

- Experience in aligning business and technical metadata.

- Maturity in managing changes in data and metadata over time (for example, establishing versions of data held in our data management systems and the surrounding metadata and using this capability to enable restatement of history in support of a need for different views of the business over time).

Skills needed to manage the data management systems that we might assess include the following:

- The ability to manage relational database data warehouses for performance (optimization and tuning), user authentication, data access, high availability, backup and recovery, and disaster recovery.

- Experience in rapidly deploying and provisioning relational databases for data marts and development projects.

- The ability to rapidly scale (shard) and manage NoSQL databases and provide data availability (replication and backups).

- Experience in rapidly deploying and provisioning NoSQL databases for development projects and for new production.

- Maturity in managing Hadoop clusters for user authentication, data access, and data availability (replication) and our experience in rapidly scaling clusters.

- Experience in rapidly deploying and provisioning Hadoop clusters for development projects and for new production.

## Application Architecture and Integration Skills

The application architecture should define how our technical solutions deliver solutions to business problems by enabling the business processes required to run the business and how these applications interact with each other. For an information architecture that is defined by many components, the interactions are enabled by key data integration components that bring all of the pieces together.

Some of the skills we might assess here include the following:

- Maturity in translating business needs and drivers into software-driven solutions that deliver information critical to making decisions.

- Experience in deploying the right mix of data discovery, business intelligence, and predictive analytics tools needed to gather information from data that can solve business problems.

- Experience in responding rapidly to fast-changing business demands using agile applications development and change management methodologies.

- Experience in linking data management systems (data warehouses, NoSQL databases, Hadoop) to data sources and transforming data as needed between systems.

- The ability to rapidly integrate data among various data management solutions.

- Maturity in delivering data in time to make critical business decisions through most appropriate data integration process (such as ETL, replication, message queuing, and so on).

- Experience in developing and deploying master data management or alternative data rationalization solutions.

## Technology Architecture Skills

The underlying technology architecture will provide a foundation for the success of our project. Skills that exist in IT are critical to defining and managing the technical architecture. For Internet of Things projects, skills that exist in our engineering organization can also be critical.

This portion of the assessment evaluates logical software, technical programming, and server, storage, networking, and communications design and management skills. The list of skills can be quite extensive and only grows in complexity with the addition of NoSQL databases, Hadoop clusters, and the Internet of Things to the information architecture. Some of the skills we could assess include the following:

- Experience in defining a logical data warehouse in the information architecture that consists of traditional data warehouses and data marts, NoSQL databases, Hadoop clusters, and data integration tools and utilities.

- Experience in defining and rapidly deploying test and development environments for key components in the information architecture (for example, the logical data warehouse and business intelligence and data discovery tools).

- Maturity of design and capacity planning skills for servers and storage enabling optimal database footprints for enterprise data warehouses and data marts and a demonstrated ability to successfully deploy and manage these platforms.

- Experience in designing, deploying, and managing servers and storage that provide optimal NoSQL database footprints.

- Experience in designing, deploying, and managing servers and storage that provide optimal Hadoop cluster footprints.

- Maturity in the ability to evaluate appliances vs. build-it-yourself systems (for example system design skills, time-to-market considerations, costs of alternatives, and flexibility).

- Experience in designing, deploying, and managing business intelligence ad hoc query and reporting and data discovery middle-tier servers.

- Experience in designing, deploying, and managing data integration solutions including ETL and data replication.

- Experience in networking together data management system servers, middle-tier servers, and other information architecture components.

- Maturity in designing and deploying data centers that meet environmental needs of the systems (such as cooling, power and floor space).

- Maturity in securing data centers.

- Experience in design and deployment of resilient data centers (such as primary and secondary sites, optimal networking between sites, and so on).

The skills needed to deploy Internet of Things footprints can extend well beyond those required for an information architecture that is focused on just an analytics footprint. For example, depending on the planned scope of the implementation, the following skills might be desirable to include in the assessment:

- Experience in designing intelligent sensors and controllers and integrating them into manufactured parts, parts assemblies, devices, and products (for example, power considerations, footprint, and ruggedization).

- Experience programming intelligent sensors and controllers (programming languages such as Java and C, event processing engines, and business rules engines).

- Experience securing intelligent sensors and controllers (software and physical).

- Maturity in provisioning intelligent sensors and controllers.

- Experience in designing communications networks for data transmission from intelligent sensors and controllers to the analytics footprint (such as Wi-Fi or other network solutions and gateways).

- Experience in securing and managing communications networks.

---

■ **Note**   The skills we described in this chapter as presented in the previous lists are meant to provide examples of areas of evaluation. You likely have many others that you believe are important to include. Depending on your focus, you might also want to perform a more detailed evaluation at this phase of the methodology for the skills we listed. For example, within a business intelligence competency center, you might want to evaluate the ability of individuals to communicate and network with others in the company and their ability to facilitate discussions, manage projects, understand data and visualization techniques, and take advantage of data and information resources. Clearly the level of skills detail you might evaluate is almost endless. Remember that the purpose during this phase is to understand whether or not the project being envisioned is feasible given current skills and, if not, what steps might be taken to overcome those gaps in skills.

---

# Addressing Skills Gaps

If your organization has all of the skills described in the previous section, chances are that you work for a very large software vendor or systems integrator, or possibly you work for a company or organization that is already building an entire footprint for an Internet of Things solution. However, the number of organizations we meet that possess all of the skills that will be required when defining and deploying Big Data and Internet of Things projects is quite small. Most face significant gaps in one or more areas.

Figure 5-3 illustrates a typical summary diagram from our skills assessment spreadsheet for an organization considering an Internet of Things project. You can see that it illustrates gaps in skills uncovered in the four architecture types we explored. The inner set of points represent survey answers regarding the current state, and the outer set represent our desired future state. The space between represents the skills gap.

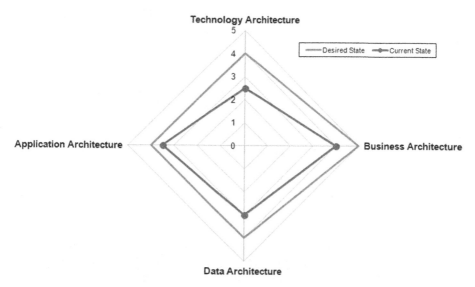

***Figure 5-3.*** *Skills gaps summary represented by a spreadsheet graph*

We can clearly see in this figure that there are skills gaps in each of the four areas. The greatest gaps appear to be in the technology architecture area. However, we would want to explore each of the individual skills gaps identified in each area as certain individual skills might be more critical than others to the success of our project.

Given the wider gap in technology architecture skills, let's next explore some of the individual skills gaps. Figure 5-4 illustrates a typical diagram created in our technical architecture skills assessment spreadsheet that represents some of the skills we previously mentioned in this chapter. Once again, the inner set of points represents survey answers regarding the current state of skills in our organization; the outer set represents our desired future state for these skills. The space between the two sets of points is the skills gap. You probably notice some of the specific skills for an Internet of Things project shown here, including sensor programming and communications network configuration and management. Hadoop and NoSQL databases are also under consideration.

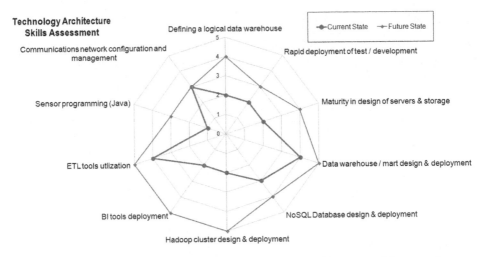

*Figure 5-4. Skills gaps in technology architecture represented by a spreadsheet graph*

From the diagrams illustrated in Figures 5-3 and 5-4, we get the notion that this organization is new to projects of this type. There are some gaps identified in traditional skills needed when defining, deploying, and managing the information architecture. There are also missing skills required for Big Data and Internet of Things projects. To be successful in implementing the project, taking steps to overcome the skills gaps will be critical.

# Delivering the News of Skills Gaps

It is now time to summarize the skills assessment information that was gathered. That information should be delivered back to all interested parties that took part in the evaluation process and to potential project sponsors in the form of a presentation and report. Delivering a presentation first in an interactive forum can help ensure that a validation process takes place where the completeness and ratings accuracy of the skills is reviewed. It will also likely spur some further discussion about missing skills and the availability of critical resources.

Generally, providing the information back to the original attendees within a week or two is a best practice and helps maintain the teamwork that was established in the original session. Given the diverse nature of the skill sets evaluated here, we will likely find it desirable to have a detailed review with each manager or management group first, validate the content, and then create an executive version for project sponsors. For example, the detailed discussion might consist of a line-by-line review of the individual skills evaluated. The executive version is usually presented in a summarized version. The diagrams illustrated earlier in this chapter could prove useful in visually calling an executive's attention to critical skills gaps.

A typical agenda for such presentations might include the following topics:

- Overall goal of the session

- Self-introduction of attendees and self-described meeting goals

- Brief review of the scope of the proposed project (as determined from early visioning and business discovery) and the critical skills required

- Discussion of current skills and gaps that were uncovered

- Validation of the accuracy of the assessment (including ratings and / or critical skills missed)

- Discussion of options available to address skills gaps and potential impact on the information architecture and the project

- Discussion of next steps and other sessions needed

After these sessions are held and further feedback is gathered, a final report should be issued that includes the suggested adjustments or at least mention of them. The content that should be included in the report includes the following:

- Who took part in the study

- How the business needs and vision drove early project definition and the skills that will be required to define, implement, and manage the proposed future state information architecture

- Breakdown of skills evaluated (by architecture type)

- Assessment of skills needed to fulfill the vision and identification of important gaps

- Proposed solutions to filling the skills gaps

- Next steps including scheduling of activities outlined in subsequent chapters of this book

As we noted near the end of the agendas for delivery of the presentations and the report, we are not simply presenting the skills gaps as a problem. We are also presenting alternatives that can address the gaps and provide solutions.

# Addressing Critical Skills Gaps

When faced with critical skills gaps, there are several approaches possible to address them. For example, the future state architecture vision could be modified to include less innovative but better understood technologies, provided the changes can still fulfill goals of the project. However, we could also find that the more challenging technologies must be included or are desirable for other long-term needs. What then?

If the missing skills are strategic to the future of the organization's success, we might evaluate training existing personnel or hiring new personnel who have those skills. Alternatively, it might be decided that the best option is to hire a systems integrator or rely on specialty consultants to provide all or some of the missing skills. The choice made will likely be decided by the expense in building vs. buying the skills and the time it will take to build the skills. We should also consider the project risk that will be introduced by using any of these approaches. Figure 5-5 illustrates how skills gaps might be remediated and the trade-offs compared. Using this diagram, the choice could become clear if the decision will be based on the importance of time to market of the solution provided by the project or if it will be based on the long-term cost to the organization.

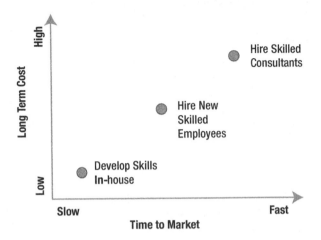

*Figure 5-5. Skills gap remediation options compared*

Other factors might enter our decision. Some employees in our company might view the project as a learning opportunity and a chance to build strategic skills within the organization. These employees might believe that shadowing hired consultants as the project proceeds could provide a solution to the skills gap. We might find that specific required skills are hard to find in consulting companies or in the market in general, leaving us with no choice other than to develop those skills internally. For example, a shortage of skilled individuals is often a challenge that must be overcome when a data scientist's skills or advanced data management solutions skills (such as those associated with Hadoop and NoSQL databases) are required.

To overcome skills shortages, some organizations that are currently defining and building Big Data and Internet of Things projects have developed innovative approaches to solving the problem. Some are working jointly with local universities to create courses of study that enable students to develop the rare, but highly in-demand skills. The organizations then formulate agreements to hire these skilled students as interns or on a permanent basis after their graduation.

■ **Note**   Internet of Things projects often include evaluations of build vs. buy for key required components in the information architecture, and these are often determined partly by the skills that may or may not be present in the organization. For example, many organizations rely on manufacturers and engineering consulting partners to provide sensors and intelligent controllers on devices. These partners might also program the devices to provide critical data. Some also rely on third parties to design and provide the communications backbones needed to transmit data from the devices to the analytics infrastructure being designed and deployed in-house.

We have just begun the skills assessment and remediation process at this phase in our methodology. Understanding the skills we need for project success will become even more critical and require more detailed analysis later as we begin to establish a roadmap to implementation. But we might find, as we take a closer look at our future state information architecture, that we have the flexibility to modify our plans and avoid requirements to add some of the identified missing skills. Even if we don't find such flexibility, at least we've already started to assemble critical information needed to develop training plans and hiring plans, and for use in obtaining implementation cost estimates from systems integrators and consultants later.

# CHAPTER 6

■ ■ ■

# Designing the Future State Information Architecture

We touched on the current state of the information architecture and a possible future state when we explored "the art of the possible" during our earlier visioning meetings. As useful as those exercises were, we only began to scratch the surface in assessing the key components in the current state. At the time, we used that limited knowledge to begin to postulate about the possible enhancements that we would need in the future. Later, we gained much more knowledge regarding the potential business use cases and prepared business information maps (BIMs) with the help of the lines of business. These requirements and desired future state BIMs will now help us further understand gaps in the current information architecture and how it must change. We will also consider the initial assessment that was made of skills present in the organization as we define the direction that the future state architecture will take.

As we begin designing the future state information architecture in this phase, we start by capturing a much more detailed view of our current state information architecture and explore the various components currently present. You will recall that we provided an initial introduction to key technologies typically deployed in Big Data and Internet of Things projects in Chapter 1. We will now look at them in the context of the architecture. During this phase, we will also capture details about the required properties of these components and how the requirements will influence the design of the future state. Once we have gathered that information, we will map the current state BIMs to the architecture to validate we have captured all of the relevant components.

In a previous phase of the methodology, we also gathered the future state BIMs. We are likely to find that we now will need new capabilities delivered in the information architecture in order to enable the envisioned changes in how the business will be run. The future state BIMs can help us determine additional data sources and data management systems needed to provide and process the required data. So, we will use these requirements to create our future state information architecture. As we introduce these new components, we will revisit various other capabilities needed in the future state footprint.

When we conclude this chapter, we will be ready to move onward in preparing a roadmap critical to securing funding for the project. We will have a much clearer picture as to the scope of the project at hand. We will understand the technology components but will also better understand additional skills that might be required to succeed.

And we will start to think about how we might deliver the future state information architecture and business solutions in phases that will best align to our current state and where we are headed.

Figure 6-1 illustrates our current phase in our methodology for success. It also outlines our focus in this chapter.

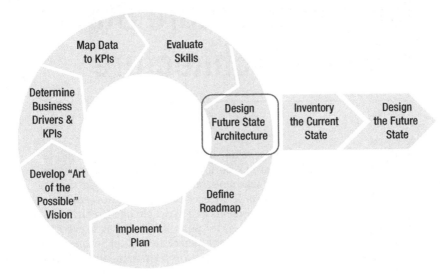

***Figure 6-1.*** *Designing the future state architecture phase in our methodology for success*

Similar to the earlier chapter that covered BIMs, this chapter on the information architecture is also divided into two sections that focus on the current state and future state. In this chapter, we will first focus on documenting the current state information architecture and validating our earlier current state BIMs. The second section will walk through defining the future state information architecture that will deliver the capabilities in alignment to the future state BIMs that were created previously.

# The Current State Information Architecture

We begin this phase by exploring the current state information architecture in enough detail so that we can understand its influence on the project we are planning. Key components that we will explore include the current data sources, data integration tools, event processing (especially common in Internet of Things deployment), data management systems for analysis, data integration tools, information access applications and tools, and other user interfaces. We will describe these partly through an illustration of a conceptual version of the architecture.

We will also evaluate the underlying security and data governance, shared services, and server and storage infrastructure. We will begin by taking an inventory of all of the components and their roles. We will also evaluate how well they fulfill current expectations. Part of this process will typically include an inventory of the vendors present and whether the current components they provide are strategic to the organization going forward.

# Data Sources

The current information architecture will include many different data sources. Some of these will be well-documented internal systems such as the ERP, supply chain, customer relationship management, and financial systems. In some industries, we will find other well-documented enterprise class systems fundamental to running the business. For example, in the financial services industry, we would find that a bank's information architecture would include a treasury management system. These internal systems usually host relational databases that are ideal for online transaction processing (OLTP) since the data is highly structured and frequent updates occur.

Other systems that provide key sources of data might be less well documented. These sometimes include external data sources that the lines of business access to populate specific departmental data marts or spreadsheets. The reports that are generated from this data might be considered critical to running the business. If we did a good job in an earlier phase of our methodology, we should have documented these in our BIMs. An example of such a data source would be an external data aggregator of marketing information. The data aggregator would be capable of providing additional broader industry data that is impossible to gather from available internal data sources.

Other existing internal or external data sources might be inventoried at this time even if they are not integrated into the broader enterprise information architecture. For example, clickstream data from internal web sites might be largely unused at this time. Sensors might be deployed on devices, but the data might not be processed today. Third parties that provide reports on customer sentiment and other characteristics important to the organization could analyze social media data. However, the third parties might not make the actual source data available.

Our understanding of all of these data sources will help us later as we look at our future BIMs and begin to understand the gaps in available data. Figure 6-2 illustrates typical data sources that serve as the starting point for our conceptual information architecture diagram.

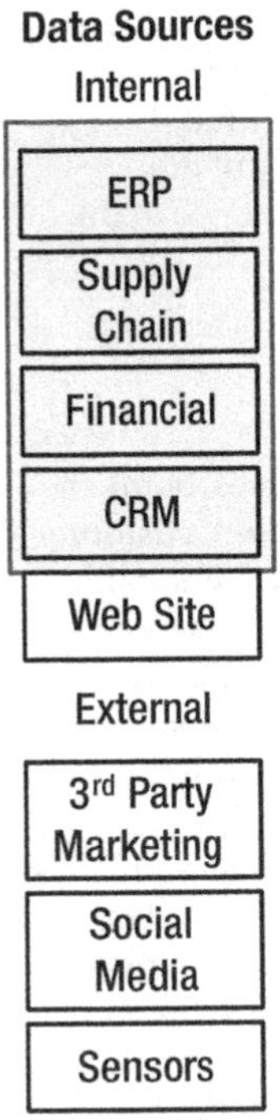

**Data Sources**

Internal

```
┌──────────────────┐
│ ┌──────────────┐ │
│ │     ERP      │ │
│ └──────────────┘ │
│ ┌──────────────┐ │
│ │    Supply    │ │
│ │    Chain     │ │
│ └──────────────┘ │
│ ┌──────────────┐ │
│ │  Financial   │ │
│ └──────────────┘ │
│ ┌──────────────┐ │
│ │     CRM      │ │
│ └──────────────┘ │
│ ┌──────────────┐ │
│ │   Web Site   │ │
│ └──────────────┘ │
└──────────────────┘
```

External

```
┌──────────────────┐
│    3rd Party     │
│    Marketing     │
├──────────────────┤
│      Social      │
│      Media       │
├──────────────────┤
│     Sensors      │
└──────────────────┘
```

*Figure 6-2.* *Data sources as illustrated in a conceptual information architecture diagram*

As we develop the conceptual diagram to represent the inventory of current data sources, we also gather other information about the levels of service that these sources provide today. We do this because these levels of service could impact our ability to deliver required business intelligence and analytic solutions in the future.

Some of the key properties we might gather include the following:

- Availability:
  - Does the source system offer 99.999% availability?
  - What level of availability is required?
  - Is there a second site for disaster recovery, and is a second site mandated?
- Recoverability:
  - What are the data archiving policies?
  - How long does it take to recover data if it has been archived?
  - Does data recovery timeframe meet business requirements?

- Performance:
  - Are there guaranteed performance levels for source systems?
  - How are source systems managed such that performance level requirements are met?
- Data Granularity:
  - What is the level of detail in the data?
  - How long is history kept in the source systems?
  - Are needed levels of detail and length of history changing over time? How are they changing?
- Data Security and Governance:
  - What sort of security for data at rest is present today in the sources? Is data encrypted?
  - What sort of access control security policies are in place? Are certain data fields redacted for certain business users?
  - Does security in place meet the requirements of an industry standard (such as HIPAA in healthcare or PCI for financial transactions)?
  - Are the same security standards maintained across the entire infrastructure? If not, how do they vary?

# Data Management Systems for Analysis

The next major portion of our information architecture that we will explore consists of the data management systems used for the analysis of data. These include our data warehouses and data marts that are commonly deployed using relational databases. Some of the data mart variations might be deployed using OLAP technology. Increasingly, Hadoop clusters are being deployed to fulfill several roles in this architecture. The combination of various types of data management systems is commonly and collectively described as a "logical data warehouse."

Today, the Hadoop clusters might be used to process streaming data from sensors and other sources. They might also serve as a landing spot of all data and serve as the desired location for predictive analytics. A Hadoop cluster can also serve as a highly parallelized and highly performant ETL engine.

Of course, the discussion of current data management systems for analysis naturally leads to consideration as to how these data management systems are populated. While ETL tools are popular choices where data quality and data rationalization is part of the process, sometimes simply an extraction and loading process is adequate, so lighter weight engines are deployed. When loading the Hadoop cluster, native utilities present, such as Flume, are sometimes used for data loading.

Figure 6-3 illustrates the progression of our information architecture conceptual diagram as we add the data management systems for analysis and key integration components. In this current state diagram, we are illustrating ETL processing as taking place independently of the Hadoop cluster.

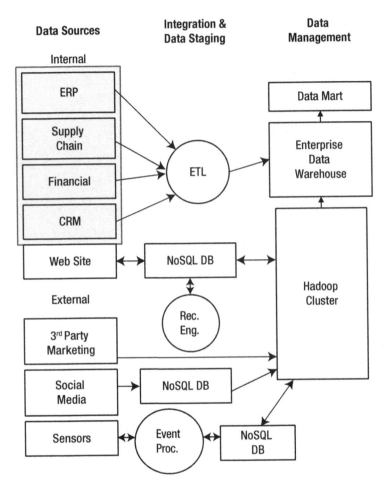

*Figure 6-3.* *Data management systems for analysis and integration components added to the conceptual diagram*

---

■ **Note** If the organization has already deployed an Internet of Things solution and is processing sensor data, the sensor data will be loaded into an analytics data management system that is likely to be a Hadoop cluster or NoSQL database. Given there is often a need for real-time actions in response to events, you might find that business rules are defined to provide this response. These would be implemented using an event-processing engine. You also see event processing illustrated as a component in Figure 6-3.

---

As we gather information about the current data warehouses, data marts, and Hadoop clusters, as well as how they are populated, we should ask questions that could include the following as we determine the properties of these systems and tools:

- Schema:

    - Is the schema in the data warehouse third normal form, star, or a hybrid?

    - What type of schema is present in the data marts?

    - If multiple data marts exist, are dimensions conformed across data marts?

- Availability:

    - What are the availability requirements for the data warehouse and / or data marts and / or Hadoop cluster and / or NoSQL databases?

    - What are the availability requirements for the network?

    - Is there a disaster recovery plan? What is it?

- Recoverability:

    - What are the data archiving policies for the various systems deployed as part of the logical data warehouse?

- Performance:

    - Are there guaranteed performance levels for any of the systems?

    - How are they managed to meet these performance levels?

    - Are in-memory databases implemented as part of the footprint? What are the impact and / or limitation of the in-memory solution(s)?

    - If Hadoop is deployed, does it leverage YARN and Spark technology?

- Data Granularity and Volume:
  - What is the level of detail in the data stored in each system?
  - How long is history kept in each of the systems?
  - What are the data volumes (raw and compressed) in each system?
  - How fast is data volume growing in the current systems?

- Data Security and Governance:
  - What level of data security is present today in the logical data warehouse?
  - Is the level of security consistent across the various systems?
  - Does the level of security meet / exceed an industry standard (such as HIPAA in healthcare)?
  - Is data encrypted in each system? Is data encrypted when transmitted over the network?
  - How is access control managed in each system?

- Data Loading:
  - How is ETL processed (for example, where do the transformations occur and how well parallelized the processing is)?
  - How often do data loads occur and how long do they take?
  - What are the data volumes moved between systems and mappings used during extractions and loads?
  - If Hadoop clusters exist, what are the data ingestion rates into Hadoop?
  - Are NoSQL databases used as front ends to the Hadoop cluster to speed data ingestion or other techniques used?

- Data Quality and Meaning:
  - Is data quality analysis and data quality improvement part of the ETL processing?
  - Is consistent metadata defined for data in the systems that make up the logical data warehouse (and is it also defined through the ETL tool)?

# Data Analysis Tools and Interfaces

A third major area we will document covers the data analysis tools and interfaces that are present. These are the tools used by business analysts and data scientists to understand the business, but also include the simpler tools used by casual business decision makers and executives. Many of the tools provide a look back at what happened, but others also enable predictive analysis of the impact that decisions will have on future business outcomes.

The simplest tools provided will report on business outcomes. They might gather data directly from OLTP source systems to show current state of the business or from any of the data management systems in our logical data warehouse.

Some of the business analysts might formulate their own reports through ad hoc queries and use ad hoc query and analysis tools to respond to changing business requirements. These tools could be pointed at any of the data management systems in our logical data warehouse. Where high data quality is required and guided drill downs are needed, they will most often be pointed at data marts or at an enterprise data warehouse structured in a hybrid schema.

Where new combinations of data need to be explored that are not supported by the schema in our data marts or data warehouses, information discovery tools can provide the means to do so. These could have access to data residing anywhere in our logical data warehouse, either directly extracting data from the various sources or accessing data in Hadoop clusters that serve as enterprise data hubs (for example, as a gathering point for all data). The data management engines underneath such tools are "schema-less," enabling free form exploration of the data.

The most advanced business analysts, statisticians, and data scientists could be using data mining and predictive analytics tools. Such tools are also typically used to analyze data in any of the data management systems in our logical data warehouse. The mathematical algorithms they provide are used to model the data and analyze patterns of outcomes. When the models are refined, they are used to indicate the likely outcome of a future event.

Any of these tools might be used to generate the intelligence needed not only to show what happened or what will happen, but also to help define the specific business processes that could be run in an automated fashion when certain results or behaviors are observed. Complementary technologies you might find deployed include Business Process Engineering Language (BPEL) scripts triggered in business intelligence tools or real-time recommendation engines that leverage predictive analytics.

Figure 6-4 illustrates a further progression of our information architecture conceptual diagram as we add the data analysis tools and interfaces.

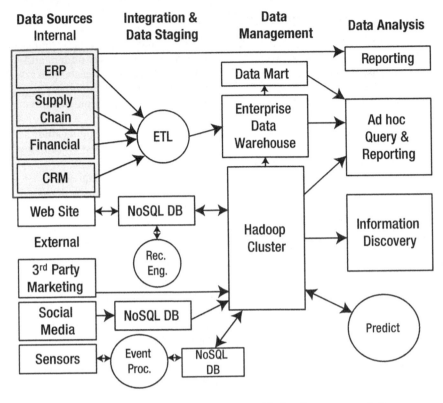

***Figure 6-4.*** *Data analysis tools and interfaces added to the conceptual diagram*

Looking further into the tools and capabilities that are present, we might ask questions that include the following as we determine their properties:

- Reports:

    - How often are the reports generated and who uses them?

    - What are the data sources and how are they accessed (for example, directly through SQL, through vendor specific options, or through generic interfaces such as Hive)?

    - How highly available must the reports be in order to meet business requirements?

    - What are the business requirements driving the need for reports?

- Ad Hoc Query and Analysis Tools:

    - Who uses the ad hoc query and analysis tools?

    - What are the data sources, how often is the data updated, how is the data accessed, and how often is it accessed?

    - How highly available must the tools and underlying data be in order to meet business requirements?

    - What are the business requirements driving the deployment of ad hoc query and analysis tools?

    - What data visualization needs must be met through the tools?

- Information Discovery Tools:

    - Who uses the information discovery tools?

    - What are the data sources and how are they accessed?

    - What drives the need for information discovery tools? How are they used?

    - What data visualization capabilities must the tools provide?

- Predictive Analytics and Data Mining Tools:

    - Who uses the predictive analytics and data mining tools?

    - What are the data sources and how are they accessed? How highly available must the tools and underlying data be to meet business requirements?

    - What are the business requirements driving the usage of these tools? How successful are these tools and analysts in meeting business requirements?

    - What data visualization capabilities must these tools provide?

- Automated Business Processes as Output:

    - What impact on the business does execution of these automated processes have?

    - What are the data sources, where can the data be accessed (for example, in data stores or streams), and how quickly does a process need to occur as the data is processed?

    - How highly available must the processes and underlying data be to meet business requirements?

# Validating Current State BIMs

If the conceptual information architecture in your organization today nearly matches Figure 6-4, you could have relatively few changes ahead at this level of abstraction. However, many of the details regarding the components present in the architecture could change as the new project is designed, developed, and implemented.

In contrast, we often find that the more common information architecture found in most organizations is missing many of the components shown in that figure. To illustrate what we typically find, let's take another look at the current state BIM gathered at the mythical LMC automobile manufacturer that was described in Chapter 4. We will then take a look at a more likely current state information architecture conceptual diagram for the company that would be behind that BIM.

You might recall that the current state BIM for the maintenance and warranty system was represented by the diagram that we reproduce here in Figure 6-5.

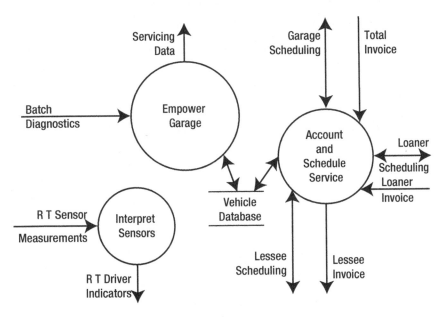

*Figure 6-5.* *Completed current state (Level 1) "LMC M&W System" BIM*

The BIM diagram illustrates an ability to download data stored in the vehicle to the vehicle database when it is in a LMC service garage (for example, as batch diagnostics). Account and service scheduling data is also stored in this database. Sensors in the vehicle cannot communicate directly in real time to the database.

We might expect the current state information architecture at LMC that matches the data flows in the current state BIM to resemble the diagram in Figure 6-6.

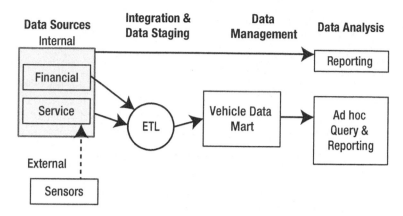

*Figure 6-6. Current state information architecture aligned to "LMC M&W System" BIM*

The architecture pictured here is one that we commonly find today where sensors are deployed in devices and products. As we revisit the future state BIM for LMC in the next major section of this chapter, you will see how the information architecture must change in order to deliver needed data and fulfill LMC's new business requirements.

## Underlying Servers and Storage

Thus far, we have focused on the software components, but we have not assessed the underlying servers, storage, and networking infrastructure currently present in the infrastructure. These components will likely also influence some of our choices as we define the future state information architecture that supports our data management systems and surrounding middleware.

As we explore these components, we will need to ask questions that might appear to overlap with some of the questions that we asked previously. However, these questions will be asked of systems, storage, and networking architects who could have different views from those expressed previously. The following questions are representative of those that we might ask:

- Servers:

    - Are current OLTP and logical data warehouse servers able to support current workload demand and easily adapt to changes?

    - Are current workloads on systems bound by CPU, memory, or throughput issues?

    - If the servers are configured to be highly available in order to meet service level agreements, how are they configured?

    - Is there a disaster recovery plan and what are those configurations for servers and storage?

- Have reference configurations been established or are engineered systems / appliances preferred?

- Are development and test systems available and are they identical to production systems?

- Storage:

  - Is the storage architecture managed separately from the servers or are engineered systems / appliances preferred?

  - Is there an information life-cycle management strategy for archiving data? What is it?

  - What backup and restore procedures are in place for each system? Are they well tested and proven reliable?

  - How flexible are storage systems to changes in workload demands and data growth?

  - What sort of RAID strategy is deployed for storage and how is this strategy impacted by service level agreements and performance requirements?

- Networking:

  - Is networking a bottleneck today within clusters or between systems in the data center, or within the internal organization?

  - What are the networking standards in place (such as Ethernet, InfiniBand, and so forth)?

  - What is the networking strategy for movement of data from sources external to the data center?

  - What service level agreements are in place for networking?

  - How is data secured when transmitted over the network?

  - Are there industry standards and / or certification levels that must be adhered to when data is transmitted (for example, HIPAA in healthcare and PCI in financial transactions)?

The level of detail that can be provided in the answers to each of these questions can be extensive and technically complicated. Furthermore, it is important to keep in mind that there can also be politically charged organizational reasons for deployment choices. For example, are engineered systems and appliances frowned upon because a storage architecture group feels threatened by their presence? Are the lines of business frustrated by the amount of time it takes IT to configure and make a new system available? The answers to these sorts of questions must also be understood if a goal is to assure that the design of the future state information architecture will be well received among key influencers and stakeholders.

---

■ **Note** A wide variation of answers to the previous questions is possible since some systems serve as data management platforms and some support middle-tier software solutions. It must be understood how critical a system and network are in running the business and whether system and network service level agreements have been negotiated and must be met. Many organizations try to standardize the design, deployment, and management of their systems and networks to ensure consistent quality of service and simplify the infrastructure. Understanding the current philosophy present in an organization is crucial to understanding how future requirements are likely to be derived and how important the stated requirements are to the future state information architecture design.

---

## Other Current State Practices

When evaluating the current state, we should also document current monitoring and administration practices used for data management and middle-tier software platforms, servers, storage, and networking components. As we design our future state information architecture, we will want to consider how to introduce new components without needlessly causing too much disruption in current approaches. The introduction of new tools and procedures will introduce additional purchase costs and a need for further skills development. We will want to understand how well received such changes might be.

We should also understand how the current organization develops and operationalizes projects. We should understand the methodology used in the organization when creating new code and testing the functionality, performance, and availability of proposed solutions. We should also understand strategies for applying patches in test environments and in production, and how new solutions are usually put into production.

# Designing the Future State

At this point, we have gathered a lot of information about the current state information architecture including accepted design, deployment, and management techniques in the organization. We applied business information maps that describe the current data flows to the underlying information architecture diagram and validated that we had an accurate representation. As described in the previous chapter, we also became aware of the skills present in our organization and through that earlier skills assessment are better able to understand our ability to take on this new project.

Now is the time to take the new business requirements that we gathered and related future state BIMs and begin to define our planned information architecture in a diagram. Once we have defined that future state, we will take a look at possible implications when introducing new data management platforms that include Hadoop and NoSQL databases and the additional technologies required in an Internet of Things project. We will also explore the early stages of operational management planning.

## The Future State BIM and Information Architecture

In Chapter 4, we described a future state BIM for LMC's more fully automated maintenance and warranty system. We began to suspect at that time that our current state information architecture was unlikely to be able to provide the underpinnings needed to deliver the required business solution.

The future state BIM for the maintenance and warranty system is reproduced in Figure 6-7. As you might recall from the earlier chapter, this diagram shows the capture of sensor data on the vehicle transmitted to a vehicle event database. Maintenance events are detected and servicing of the vehicle is scheduled in an optimized manner. All events are captured in a vehicle log database for later processing when the vehicle is in for servicing. A variety of KPIs are reported that describe how maintenance was scheduled, when it occurred, and what the outcome was. As indicated earlier in Figure 6-5, the maintenance tasks and accounting data is also stored in the vehicle database.

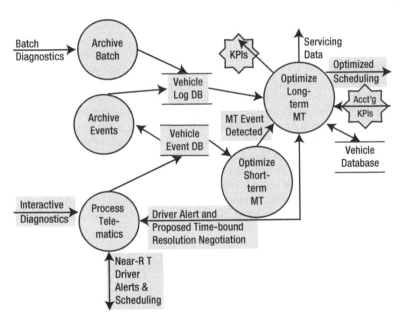

***Figure 6-7.*** *Future state "LMC M&W System" BIM*

The future state information architecture diagram we develop must produce a design that aligns to the data flows and processes represented in the future state BIM. An example of how our future state information architecture diagram might appear is illustrated in Figure 6-8.

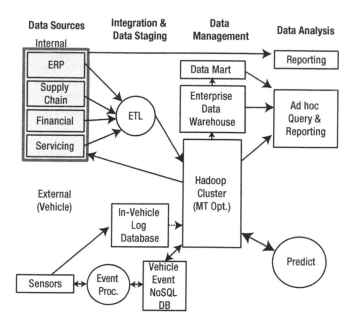

***Figure 6-8.*** *Future state "LMC M&W System" information architecture*

The difference between the future state diagram in Figure 6-8 and the earlier current state diagram in Figure 6-6 is striking. Many new technology components are added including NoSQL databases, event processing, Hadoop clusters, and predictive analytics. ETL processing for the transactional data sources is shown as being routed through the Hadoop cluster prior to loading into the data warehouse. Because of the substantial changes that are proposed in the information architecture, there will be many design, deployment, maintenance, and skills considerations that LMC must face as it transitions to the future state.

## Broad Future State Considerations

Many organizations face challenges similar to those described in our mythical LMC example when defining a revised information architecture that includes Big Data and the Internet of Things components. Because of the added complexity that is introduced, a cloud-based deployment strategy is sometimes considered as an alternative. Early test and development of future state components often takes place in the public cloud today. However, as testing and development winds down, a hard look is usually taken at the data volumes that already exist in on-premise solutions and data volumes that could exist in the cloud. If data volumes and data movement requirements are substantial between cloud-based and on-premise solutions, performance demands placed on networks could present a challenge. As a result, during this transition period, evaluations sometimes take place regarding re-hosting all of the platforms in the cloud, deploying the architecture as a hybrid cloud and on-premise solution, or hosting data gathered from the new data sources as part of a broader on-premise information architecture. Of course, the cost and flexibility trade-offs of each of these approaches are compared as well.

We must also re-evaluate how the approaches used in the current state deployment that we have documented will apply in our future state. For example, the volume of data in the data warehouse could grow substantially as we gather more data from a wider variety of sources. Management, tuning, and backup procedures that were formerly seen as adequate might prove to be too cumbersome to handle the new requirements. Existing reporting and ad hoc query solutions might also be inadequate when addressing the new business problems to be solved. So our focus should extend beyond the new platforms to also include necessary changes in the existing footprint.

---

■ **Note**  One example of the potential change to the existing footprint is consideration of moving ETL processing to the Hadoop engine, as we pictured in Figure 6-8. The highly parallel nature of Hadoop makes it ideal for processing the complex data transformations that often occur in a logical data warehouse.

---

Data governance is a topic that requires special attention during this phase. We should take another look at the level of data quality needed across the entire footprint. Data quality can be less important where predictive analytics is applied to a huge data set (since given the statistical techniques used in the problem solving, low quality data is likely to be eliminated as noise). In comparison, reporting and ad hoc query solutions can present radically different results when suspect data is present. When using such tools, only trusted data is normally desired. Hence, in the information architecture that our future state diagram represents, the enterprise data warehouse will remain the historic database of record. A master data management (MDM) strategy is more likely to be put into practice for the enterprise data warehouse and across data marts if pursued as part of our future state strategy.

Consistency of data across the entire revised footprint will be a challenge if the intent is to mix data from various sources in reports or during ad hoc queries. Hence, business analysts must understand data lineage so that they can track the transitions in the data that occur as it moves among this blend of data management systems. Consistent metadata will also be extremely useful and should exist across ad hoc query and reporting tools and ETL tools, as well as where it is defined in data management systems.

## Hadoop Considerations

As described in our mythical example, the inclusion of Hadoop in the future state information architecture for LMC makes tremendous sense based on the characteristics of the data that is to be transmitted by the telematics system when it is gathered from sensors. Though LMC's current skills gap could be a challenge, developing those skills appears to be a wise investment. However, Hadoop introduces other specific considerations that should be top of mind.

Data volumes coming from sensors or other streaming data sources (such as social media feeds) can be huge and are likely to grow. Ingestion of data into the Hadoop cluster needs to be highly scalable. To land such data volumes, a common architecture today is to front-end the Hadoop cluster with NoSQL databases that can be scaled rapidly as needed. A possible alternative (being explored in many organizations at the time this book was first published) is to use Kafka's publish and subscribe model in Hadoop.

Hadoop presents unique availability challenges. The primary means of assuring data availability today is to triple duplicate the data. Often, one of the versions of the data is duplicated to another site to provide a means for disaster recovery. Special considerations of how to handle data if a cluster goes offline or communications interfaces fail must be planned for. Furthermore, data duplication is usually seen as the only viable strategy for having a backup copy elsewhere. Data volumes tend to be so immense that most consider a restore capability for data stored in Hadoop clusters to be impractical.

Sizing a Hadoop cluster configuration can also be challenging. The nodes in the cluster consist of CPUs, memory, and storage. Required raw data storage and duplication tend to drive how much of the sizing activity takes place today. Though it might be tempting to size a cluster by first computing the raw data storage requirements and then comparing that value to storage capacity provided by a number of nodes (also dividing it by the data duplication factor), such an approach can be a recipe for failure.

It is important to remember that a cluster must also provide temporary and working space for MapReduce jobs and other workloads. A common best practice is to size the cluster usable capacity to be around 65 to 70 percent of raw storage. For example, if a cluster has 600 TB of raw capacity with a duplication factor of three, we would compute 65% of 200 TB (the remaining storage after accounting for duplication) and figure that 130 TB of data could be stored. Of course, as most Hadoop clusters support various levels of compression with rates varying based on data types, the 130 TB of data storage is simply an initial but conservative estimate.

Since storage capacity grows with the addition of nodes, when capacity is reached, many simply add more nodes to a cluster. Additional nodes are also desirable when sizing for performance. For example, spreading the data over more disks and more nodes reduces contention.

The need for proper memory sizing became more significant as in-memory technologies (for example, Spark) assumed an important role in processing the data in Hadoop. A growing number of information discovery and data access tools now rely on Spark. In-memory processing capabilities are driving a desire for larger memory capacities in individual nodes (often configured with 128 to 512 GB as this book was being published) and also lead to the addition of more nodes to clusters. Since in-memory technologies utilize all available cores in CPUs, the scaling of CPU cores that occurs with the addition of more nodes is also helpful.

Sometimes overlooked, assuring adequate interconnect speed among the nodes in a cluster enables flexibility for handling a variety workloads. As production-level performance demands become more challenging, many choose to re-evaluate the trade-offs of using commodity servers and networks vs. engineered systems and appliances that are pre-configured and contain large memory configurations and scalable high-speed interconnects.

---

■ **Note**  For optimal performance, nodes in a Hadoop cluster addressing a specific workload should be physically close to each other. While this is intuitively obvious when deploying an on-premise solution (for example, typically one would network new nodes into the cluster physically adjacent to existing nodes), this is not always a given in public, cloud-based deployment scenarios where provisioning of new nodes can be automated. When you seek to provision additional adjacent nodes, you could find that no adjacent nodes or servers are available. If this is a production cluster with challenging workload demands that you wish to add nodes to, your cloud provider might recommend a re-platforming of the entire cluster to a different location in their facility or to a different facility where enough adjacent nodes are available.

---

Securing data in the Hadoop cluster also requires consideration. Recent improvements in security capabilities in Hadoop should help you define a secure environment. Securing access control is possible using Kerberos. Of course, it will be helpful to you if Kerberos is also a currently supported and deployed as a security mechanism for other proposed and existing data management systems. Alternatively, you might decide to pass SQL queries through your data warehouse or other query engine to the Hadoop cluster using one of the vendor solutions that exists today. For database-centric solutions, you can simply leverage the security model present in the relational database, which greatly simplifies how you deploy and maintain data access control.

You should also consider the security of data at rest in the Hadoop cluster and data in motion within the interconnect and the network connections to other data management systems. Encryption capabilities in current Hadoop distributions can enable you to secure this data. To track how people are accessing the data and what they are doing with it, you will likely want to include auditing tools as part of your future state information architecture.

## Internet of Things Considerations

Our LMC future state information architecture diagram in Figure 6-8 that serves as an example in this chapter includes data that is gathered from sensors installed in vehicles. The data is transmitted to a vehicle event NoSQL database and further processing takes place in Hadoop. This scenario is consistent with what many would define as an Internet of Things footprint. When designing the future state information architecture for such a project, the footprint extends beyond traditional data management components to sensors that feature software used for data gathering and intelligent actions. Sensor and device management, security, and event processing software are sometimes referred to as middleware components and provide critical functions. Design of the network linking these components together (including required gateways) is also part of the definition of the architecture.

The intelligent sensors pictured in the diagram are sometimes pre-programmed by the sensor supplier or could require custom programming. Most intelligent sensors support programming in Java, though sometimes other languages are supported or required.

Sensor and device management software is needed to register and manage the devices. The registration process enables an inventory of the devices to be kept. Sensor conditions are monitored and diagnostics can be applied when problems are detected. The sensor and device management software can also manage updates and provision software to the devices and sensors. Given its role, it should also have access to an identity directory to assure that only authorized updates can be pushed to the devices. In our example, LMC would not want a rogue command to shut down a lessee's vehicle and possibly create a dangerous situation.

Event processing enables intelligent action to be taken immediately when sensor readings detect certain conditions. The business rules that are applied are pre-programmed and usually match best practices that might have been manually applied previously when certain conditions existed. They are typically directly applied to data streams (using languages such as Java) or to data residing in NoSQL database engines. In our example, event processing is used to trigger an appointment for servicing the vehicle at LMC Service when a vehicle part fails, is about to fail, or other maintenance is required.

---

■ **Note**   Data transmitted from telematics systems and via sensors often is first sent to a "cloud" for staging. It might remain in the cloud if other key data and platforms for analysis are hosted here, or it might be transmitted to an on-premise platform. As noted earlier in this chapter, the required data volumes and network bandwidth will help us define the right architecture to deploy.

---

Given the complexity of creating, managing, and maintaining a network, most organizations establish partnerships with providers of mobile communications. However, close coordination regarding specifications for network availability, strength, and resiliency are critical as is joint engineering to assure communications between sensors and the data management systems where processing occurs is maintained.

As noted in Chapter 1, a variety of open source and standards organizations and consortia are establishing data, communications, device connectivity, device management, and architecture standards. These are sometimes seen as introducing further complexity into selecting the platforms and communications networks to be deployed. We illustrate where communications and data standards play key roles in describing and defining an intelligent device in Figure 6-9.

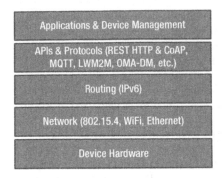

*Figure 6-9.* *Typical Internet of Things device communications and data standards*

Select industry groups are also defining and embracing certain standards. As you define the future state, understanding and embracing these standards can be critical to simplifying the initial deployment and ensuring that future support and ongoing modifications of the footprint will be possible. Please refer to Appendix B in this book for a reference list of standards, open source projects, and consortia that you might explore and consider.

## Early Operational Planning

Although we will revisit operationalizing our information architecture in Chapter 8 when we describe implementing our project, it is not too early at this phase of our design work to begin considering how we will do that. Failure to do so early and often could result in an innovative prototype that cannot be supported in production. We have already begun to touch on some aspects of placing our design into operation in this chapter. Let's briefly take a look at how we might assess the tasks at hand.

Key tasks in operationalizing the solution will include day-to-day operations, monitoring, change management, applications release management, tuning, patching, hardware and software updates, and data protection. To first assess who will perform these tasks, a RACI table or diagram is often prepared. *RACI* is an abbreviation for *responsible, accountable, consulted,* and *informed.* You will see in Figure 6-10 that we've mapped RACI to the tasks and to a set of individuals (stakeholder and line of business, business analyst and data scientist, system administrator, database administrator (DBA), storage administrator, network administrator, application developer, and IT managers).

| Activity / Task | Stakeholder & LOB | Business Analyst / Data Scientist | System Admin. | Database Admin. | Storage Admin. | Network Admin. | Application Developer | IT Managers |
|---|---|---|---|---|---|---|---|---|
| Day-to-Day Operations | I | | R | R | R | R | | A |
| Monitoring | I | | R | R | R | R | | A |
| Change Management | I | I | R | R | R | I | R | A |
| Application Release Management | C | R / I | I | R | I | I | R | I |
| Performance Tuning | C | I | R | R | R | C | R / I | I |
| Patching | I | | R | R | | I | R | A |
| Hardware / Software Updates | C | I | R | C | I | | R | A |
| Data Protection | I | | R | R | R | | | A |

R = Responsible   A = Accountable   C = Consulted   I = Informed

*Figure 6-10.* *RACI table: Operationalizing the future state information architecture*

Preparing a RACI table can be useful to us in several ways. It can help us define those who will be responsible for critical tasks and, therefore, the most likely individuals to target for skills development and training. It can also help us understand the kinds of operational management tools required. Those shown as being held accountable will likely pursue the right skills development for the operations staff and assure that the right tools and strategies are put in place for managing the infrastructure. The table can also help us define where ongoing communications will be required in alignment with the various tasks.

---

■ **Note**   RACI tables are unique to organizations and are somewhat dependent on the components and strategies present in the organization's information architecture. While the example table in this chapter denotes typical roles and responsibilities, the version that you prepare for the organization that you are working with will likely require many changes. You might also create RACI tables that provide more detail than in the illustrated example by defining detailed tasks behind each of the broad tasks shown in Figure 6-10. For example, you might create a RACI table for patching that focuses just on the individuals that manage that entire process.

---

The tools and techniques used in managing Hadoop and Internet of Things platforms are undergoing a rapid evolution. This change is being driven by the introduction of new features and functions in the platforms and the need to simplify management of already existing capabilities. While specialized tools exist from individual vendors to manage their platforms, many organizations also explore extending the tools they already have as they introduce Hadoop and Internet of Things platforms into their information architecture.

# The Right Time to Define a Roadmap

An extensive list of current technical design, deployment, management, and support documentation and plans for the future are gathered and created through the activities we described in this chapter. When we have gathered all of this information, we should have a much clearer picture of what the future state information architecture will look like. For the first time, we are likely to have gained a perspective as to the degree of organizational change that could be ahead and how complex the task at hand will be.

Some participants will likely want to refine the technical plan in fine detail at this point. However, in this phase of the methodology, we still don't have a funded project. What is most critical to obtaining that funding will be getting buy-in from key stakeholders and sponsors based on the potential return on investment that the project will provide. In computing that return, we will need to understand how much the project will likely cost. As we explore the costs, the biggest portion will not come from buying or maintaining the technology components. Rather, the greatest costs will come from the building of the solution.

Whether we will now feel ready to move forward into the next phase is likely determined by how well we believe we can get our arms around the total cost of the project. In the next phase of our methodology for success, the cost numbers we gather will not need to be exact, but neither can they later prove to be an order of magnitude different from the amount we will request in our funding and project proposal. We need to have enough of our solution defined to be confident that we can put together a realistic assessment of the benefits, costs, and risks in pursuing this project in the next phase of our methodology.

# CHAPTER 7

■ ■ ■

# Defining an Initial Plan and Roadmap

The design of our future state information architecture was described in Chapter 6. Given how it was developed, it should provide a solution that aligns with the organization's business requirements. For now, we will put the technology design work aside. Based on our earlier assessment of skills, we believe we can cover any skills gaps and successfully deploy and maintain this architecture (though we must now explore this further). Our sponsors are pleased so far. But, we still don't fully understand what it will cost to implement this design nor do we understand how quickly we will begin delivering value to our business sponsors and the organization. This likely has our sponsors concerned since we are reaching the point at which we should gain buy-in from senior executives and obtain project funding.

This chapter describes how we define our initial plan for deployment of the future state architecture. We will then describe how to lay out a roadmap that will answer the questions we just raised and address other concerns. The roadmap will target the organization's executives and provide enough information to gain funding for the project. In other words, this is the phase at which the project will gain a real commitment to move forward.

By following our methodology thus far, we should have laid the groundwork for success in gaining such a commitment. Our thoroughness in gathering business input throughout the process has likely been noticed by potential sources of project funding. Our attention to technical detail should also have our IT proponents excited and optimistic. Thus, coming forth with a roadmap and a request for budgeting should be expected, rather than a surprise.

As we prepare the roadmap, we will incorporate all of the relevant background information that we gathered. Now that we have better defined the future state information architecture, we are ready to revisit the skills that we believe are required in greater depth. We will also revalidate the business priorities and map these to project phases. Then we will be ready to figure out the cost of a phased implementation, with skills required possibly provided by internal employees and new hires or by consultants and systems integrators who will be engaged. We will also gather pricing of hardware, software, networking, other components, and ongoing support so that we can put together a more accurate business case.

We will cover all of these topics as we describe assembling and delivering the roadmap presentation to a broad executive audience later in this chapter. We will also describe what we can do if "no" is the answer to our first request for funding, and how to transition to the implementation phase if the answer is "yes."

Figure 7-1 illustrates the current phase in our methodology for success as well as the previous phases that brought us to this point. Information gathered in all of the earlier phases will contribute to the current phase. Though the implementation phase of the project plan is next, getting past this phase can be the most difficult step for a variety of reasons, which we will describe in this chapter. Of course, the most significant reason is that the organization faces the decision to commit to spending money on the project and that decision will have political and financial budgeting implications.

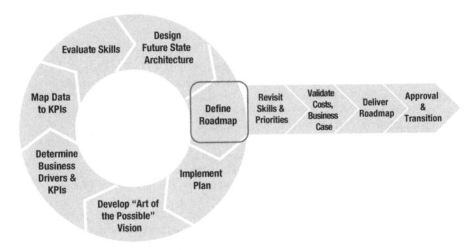

*Figure 7-1. Roadmap definition phase in our methodology for success*

We will begin by taking another look another look at the skills required. Recall that in Chapter 3 we described an omni-channel strategy that is becoming popular among retail companies as well as any organization that has both physical and web-based interactions with consumers and customers. We will use an omni-channel initiative as our project example in this chapter. For this type of project, implementation of our future state information architecture can clearly be labeled as a Big Data and Internet of Things project and it will require many new skills.

# Revisiting Earlier Findings

In Chapter 5, we outlined a broad list of possible skills we might need in order to successfully define, deploy, and manage our project. Our skills assessment helped to identify where we had significant gaps to overcome. Now it is the time to look at these skills requirements in detail so that we can better understand the costs of developing and acquiring such skills.

We also began to prioritize how the rollout of our business solutions could take place much earlier and discussed this process in Chapter 3. Now we can take another look at our earlier prioritized list with our skills gap analysis in mind and further refine the order in which we will deliver the project's business solutions. For example, we might want to move portions of the project into the delivery schedule sooner if we possess all of the needed skills and defer other phases of the project where we will need time to develop or acquire the skills.

# Refining Our Skills Assessment

We will continue to use the skills assessment methods here that we described previously in Chapter 5. We documented the skills required, the current state of skills, and the desired future state in a spreadsheet that was illustrated in Chapter 5. But now, we will break the skills categories we previously outlined in the spreadsheet into more detailed individual skills as we zero in on possible skills gaps that we will need to fill. Understanding the skills required at this level of detail will be critical when we develop a Request for Proposal (RFP) later in the chapter and seek to gather more accurate cost estimates for our project implementation and ongoing management.

Our future state information architecture is likely to require a mixture of skills we already possess and skills that might be scarce in our organization. We should re-evaluate all of the required skills in more detail. In this chapter, we'll focus on some of the skills that might prove to be especially scarce in a Big Data and Internet of Things project.

For example, since the future state information architecture that we defined for our omni-channel project requires a Hadoop cluster, we have found that we now need to further investigate the presence of those skills in our organization. There are three important types of individuals who must possess such skills and who will be critical to the success of the project: Hadoop administrators Hadoop developers, and data scientists.

Hadoop administrators play a key role in the definition and deployment of our Big Data platform. Their skills are also critical for ongoing successful management of the system. In most organizations, these individuals will reside in IT. Some of the specific skills they should possess and that we should evaluate include an understanding of and expertise in the following:

- Hadoop internals including MapReduce, HDFS, and YARN

- Server and storage infrastructure sizing for clusters and / or deployment to cloud infrastructures

- Cluster services configuration, monitoring, and troubleshooting

- Resource scheduling and performance tuning (to meet service level agreements)

- Hadoop cluster availability (through elimination of single points of failure during design of the Hadoop software and infrastructure)

- Data availability (through data duplication within the cluster and / or to secondary clusters), backup, and recovery

- Hadoop data governance and security (including authorization and access control, authentication, encryption of data, and auditing)

- Hadoop loading (including Flume for loading streaming data into Hadoop and Sqoop for loading data from relational databases, and other vendor loading options)

- Deploying, configuring, and managing HBase

A gap analysis for this role, using the spreadsheet that we described in Chapter 5, could produce a diagram like the illustration shown in Figure 7-2. The figure clearly illustrates our assessment of current skills (the inner linked set of points) and what we believe the future state skills should be (the outer linked set of points). The space between indicates a gap that we need to fill. Recall that our analysis takes into account whether the skills are present at all and how widespread those skills might be. This diagram illustrates that we have some of the skills, but that we need more individuals possessing these skills for our project to succeed.

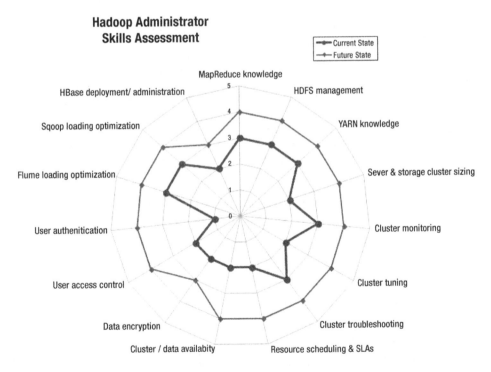

*Figure 7-2.* *Gap analysis diagram for an organization evaluating Hadoop administrators*

---

■ **Note** In organizations new to Hadoop, the administration skills sometimes primarily exist in research and development environments or in innovation centers. If the organization does not yet have a Hadoop production environment, some of the "enterprise ready" skills needed might not yet be fully developed or understood (such as the need for high availability, advanced security, and defined service levels).

---

A second set of skills we should further assess would belong to the Hadoop developers. The Hadoop developers must possess skills required in building the software that analyzes data transmitted from sensors and from other streaming data sources. These developers also generally reside within IT at most organizations. They should possess an understanding of and expertise in the following:

- The Hadoop architecture and ecosystem

- MapReduce and other capabilities and features used in analyzing data in Hadoop

- Data science (further described below)

- Spark (for speeding performance)

- HBase as a columnar data store for providing flexible data access

The data scientists sometimes reside in IT, but more commonly work alongside business analysts in innovation centers or the lines of business. The skills they should possess include an understanding and expertise in the following:

- Programming for data access and manipulation (such as the ability to program using languages such as Java and Python and Hadoop features such as MapReduce, Pig, and Oozie, and the ability to optimize applications that are developed for YARN, Spark, and so on)

- Using information discovery tools to understand data characteristics

- Understanding how to use business intelligence tools and SQL to access data in Hadoop (via Hive, Impala, or other vendor interfaces)

- Using advanced analytics and machine learning tools and techniques (R, SAS, Mahout, and so forth)

As you can see, the list of skills discussed here are quite a bit longer and a lot more detailed than those we presented earlier in this book. It is this level of detail that starts to describe the kinds of skilled people that we will need for our projects to succeed and will form a basis for writing job descriptions and identifying desired certification levels for the individuals. This level of detail is also a better indication of the type and amount of skills development and training that might be required for existing personnel.

Where scalable NoSQL databases are used as ingestion engines to front-end the Hadoop cluster, we will also need to further evaluate our skills in configuring, managing, and programming these platforms. Typical administration skills needed include installation, configuration of replication and fault tolerance, sharding, indexing, performance tuning, monitoring, and backup and recovery skills. In addition to understanding how to code in languages such as Java, Python, and C#, desirable NoSQL database programming skills that will be required include working with the language drivers, JSON, dynamic schema design in collections, indexing, and insertion and querying of data.

If part of the project relies on building out other portions of an Internet of Things footprint, including intelligent sensors and controllers and communications backbones, then the skills required become even more diverse and some of the skills will reside in engineering roles, not in IT. For example, intelligent sensor development skills often reside in teams of electrical engineers and software engineers.

Typical skills the electrical engineers will need to possess when engaged in sensor development include an expertise in and an understanding of the following:

- Microchip design

- Analog and digital circuit design

- Power supply design and battery charging

- Audio circuit design

- Printed circuit board (PCB) design, layout, and prototyping

- Radio frequency (RF) design

- Wireless technology (such as WiFi, Bluetooth, Bluetooth low energy, cellular, GPS, and Near Field Communication [NFC])

- User interface design (such as LEDs, buttons, and switches)

- Hardware and sensor integration

- RFID integration

- Root cause failure analysis

- Component analysis and optimization

The software engineers taking part in intelligent sensor and controller design work will likely need to possess skills in the following:

- Hardware device drivers

- Wireless modules (such as WiFi, Bluetooth, Bluetooth low energy, Cellular, GPS, RFID, and NFC)

- Control logic

- Power management

- Security

- Diagnostics

- Data acquisition

- User interface design

- Operating systems (Linux, iOS, Android) and embedded firmware

The skills needed to establish and maintain a communications backbone between the sensors and analytic infrastructure reside in network programming specialists and network designers and administrators. The network programming specialists will work with the business analysts and data scientists to understand the impact of application workloads on the network. They also work with the network designers and administrators who have deep routing, switching, and network availability knowledge, as well as knowledge and experience in managing and securing the network.

## Another Look at Project Priorities

In Chapter 3, we described a prioritization process for choosing projects and the priorities within projects based on the strategic impact and value, and also based on the risk and degree of complexity that was present. At this point, we are beginning to better understand the risk to project success caused by skills gaps in our organization. We might choose to re-evaluate our project choices at this time and choose to instead pursue a project with better alignment to our skills.

However, the strategic impact and value of the project and choices we have made could be so great that it makes sense to take the risk, especially if the skills identified as missing would impact our ability to pursue other projects that also are of high value. The challenge we face is to succeed early and often when deploying and managing a project where some of the key skills are missing.

One technique commonly used is to break the project into phases that incrementally deliver business value. The project begins with phases that can be more easily accomplished while leaving the more difficult phases for later. For example, we might be engaged in a project where building a fraud detection solution is of high priority. Gathering all of the data together in a Hadoop cluster that would serve as a data reservoir and performing advanced analytics there could make tremendous sense. But if we lack current skills to deploy that solution in a timely manner, we might take another approach. Instead, we might initially solve the business problem with a traditional data warehouse using skills we already have in-house, while building Hadoop skills behind the scenes as we start deployment of a data reservoir with plans to eventually transition the workload to it.

Remember that a positive return on investment can be significant no matter what technology solution is deployed, and even if it only provides an interim solution. If we obtain the desired business results sooner and can demonstrate a positive ROI, it is not a waste of money even if some of the technology is later replaced.

For Internet of Things projects, the design and coding of intelligent sensors and the setup and management of the communications backbone might be new and very different skills from those possessed in an organization. Some organizations choose to focus on just the analytics aspects of the solution and rely on others to build out intelligent sensors and provide communications networks.

Having some notion as to how a project might be deployed in phases becomes very important as we build our business case in the next step in our process. We need to gather detailed technology and implementation costs. While we can gather all of the data ourselves, there are a host of vendors, consultants, and systems integrators that can provide us with help by estimating costs based on their experience.

# A Defensible Business Case

A defensible business case must contain believable benefits. You might recall that in Chapter 3 we took a pragmatic approach. Our goal at that early phase in our methodology was to determine if we might have a viable business case. Now we are ready to confirm that we do.

At the earlier phase, we also had a very preliminary estimate as to costs. We were not very sure of what the future state information architecture would look like. And we had no idea as to the cost implications that a skills gap might cause. Now we are ready to put believable costs into the equation. Consequently, we next explore how we might go about obtaining the real costs.

## Obtaining a Real Estimate of Costs

A common method used to gain a picture of the true costs for the implementation of a project and later ongoing support is to engage vendors and consulting partners in providing cost estimates. Often, this is done through a Request for Proposal (RFP) process. A typical RFP contains questions regarding servers, storage, software, networking, and implementation and management capabilities, and it seeks the costs of providing and supporting those.

The following are some of the key areas typically covered in a Big Data and Internet of Things RFP:

- Cluster / node (server and storage) technical footprint including CPUs, memory, storage capacity, performance, and environmental details as well as pricing and support costs

- Cluster interconnect technical details, as well as pricing and support costs

- Data management system software technical details including performance and scalability, manageability, availability, functional features (support of data types, analytic capabilities) and security features (for access management, encryption, and auditing), as well as pricing and support costs

- Data integration software technical details including performance and scalability, data management systems and sources supported, and capabilities for enabling data lineage and data quality initiatives, as well as pricing and support costs

- Data discovery, business intelligence, and advanced analytics technical details including performance and scalability, data management systems and sources supported, as well as pricing and support costs

- Intelligent sensor and controller technical details including programmable capabilities, communications protocols supported, security features, environmental considerations, as well as pricing and support costs

- Network backbone technical details (used to link intelligent sensors and controllers to our analytics infrastructure) including communications protocols supported, security features, as well as pricing and support costs

- Technical training curriculum for deployment and management of the above, including costs of training modules

- Business training curriculum for optimal usage of the data that the future state information architecture will provide, including costs of training modules

- Implementation costs for the above technology components as linked to project phases and solutions delivered

- Any other ongoing management, support, or Cloud services costs that might be proposed

As the RFPs are typically sent to many potential vendors and implementation partners, a variety of solutions and options could be proposed, sometimes providing very different capabilities at very different price points. Providing the right level of detail about our requirements and guiding the responders in providing consistent information requires upfront planning. We want to compare apples to apples to apples, not apples to accordions to automobiles.

---

■ **Note**    We explicitly described this as a Request for Proposal (RFP) process, not as a Request for Information (RFI) process. Our experience is that the RFI process is often an attempt by an organization to obtain an education from its vendors, but it sometimes occurs without linkage to any real or compelling business initiatives. The answers received back frequently miss the mark because there is no mark described in the RFI. In this chapter, we instead outlined a RFP process that is linked to the work we described earlier in this book. That work resulted in clearly identifiable key business initiatives. Due to our earlier efforts, we also have an understanding of our skills gaps and what the future state architecture is likely to look like. We are relying on our partners to fill in the blanks by suggesting specific products that map to our architecture and talented individuals who can fill the skills gaps. As they provide that information, they will also provide us with the costs associated with their recommendations.

---

As we review the RFPs, we should be gaining confidence in our ability to deploy a successful project, and we should gain a much clearer picture as to how the pieces will come together as well as the true costs of the project. Now, it is time to take this input and revise our business case.

## Revising the Business Case

In Chapter 3, we created a business case based on an initial estimate of our project's total cost of ownership (TCO), IT value, and business value. Among values included in our TCO calculations were hardware acquisition and support costs, software acquisition and support costs, environmental (floor space, power, and cooling) costs, and implementation, migration, and training costs. During that phase of the assessment, the costs we plugged into our spreadsheet were very rough estimates based on a preliminary vision of the future state information architecture.

Upon completion of the RFP process, we will have a variety of proposals that we can map to a more detailed future state information architecture that we created as described in Chapter 6. That design was based on the real business requirements that we uncovered along the way. We are likely to find in the RFP responses suggested improvements to our design, and we might choose to incorporate those. Most importantly, we now have some real cost estimates to include in our TCO calculation. Using the same computational model introduced previously, we now refine our TCO calculation.

The RFP responses might also provide us with additional IT and business value propositions that we didn't initially consider. These can be very important, especially if our initial costs were greatly underestimated and the revised costs now make our business case appear to be questionable. Other trade-offs we might consider to reduce costs include taking a closer look at costs of on-premise computing vs. cloud-based solutions and costs of developing skills in-house vs. hiring consultants and systems integrators.

In most organizations, the CFO and other business executives responsible for funding will ask us about where the cost figures came from. Providing this level of diligence and detail will go a long way toward gaining confidence in the numbers and support for the project.

## Defining the Roadmap

We are now ready to create an initial plan and roadmap to implementation. The initial plan we describe here will target the IT organization and provide details containing the current state and future state information architecture, business drivers behind the future state architecture, and project phases, as well as how to provide needed skills, costs, and risks to project success. This detailed information will augment the information we will provide in a roadmap to implementation presentation.

The roadmap to implementation will be prepared for a very different audience. Our organization's executives and sponsors are unlikely to be concerned with all of the detailed information we gathered. They simply want to understand why the project is being proposed, what the value of the project is to the organization, where there might be risk and how we can overcome it, how much it will cost, when it will begin delivering business value, and how much value they can expect. Hence, the roadmap to implementation presentation is mostly business-oriented and very much to the point with just enough technical information to accurately convey the full scope of the project.

# An Initial Plan for IT

Our initial plan for IT will provide a level of detail that a technical audience can appreciate. The purpose in gathering our discovery content together is to explain why and how the future state information architecture design was arrived at and how the implementation will proceed. This document will be an important reference later if plans are modified during the project implementation. It also will help answer detailed questions raised during our roadmap to implementation presentation that we will shortly describe.

Our experience is that the initial plan is usually created in two forms. A detailed document is prepared that gathers all of our work on the project so far. A presentation is also assembled that is used for detailed briefings about what the plan contains and for discussions with technical audiences.

A typical initial project plan document contains the following information:

- A discussion of the project's history so far (when it began, the process that was used in developing the project plan, and who was engaged in the process)

- Our initial visioning outcome (potential business drivers discovered and the impact on current state architecture)

- Key business drivers and KPIs (including critical success factors, key measures identified, and business priorities)

- Data source mappings and the analysis necessary to deliver information aligned to our key business drivers (Business Information Maps)

- Detailed current state and the eventual future state information architecture diagrams

- Project phases and evolution of information architecture over time (including skills needed, ways skills gaps will be filled, and cost estimates of project phases)

- An evolving business case linked to evolving information architecture project phases

- Risks to the project's success

- Next steps upon project approval

The supporting diagrams will make their way into the technical presentation, as will summaries of key points that should be highlighted (such as the history, business drivers, skills coverage plans, and risk / risk mitigation). A diagram similar to Figure 7-3 might be used to illustrate and summarize the major project phases from an IT perspective. The figure shows when testing and development begins and when major tasks transition into production. Other more detailed diagrams would be provided offering a drill-down into steps behind each of these tasks.

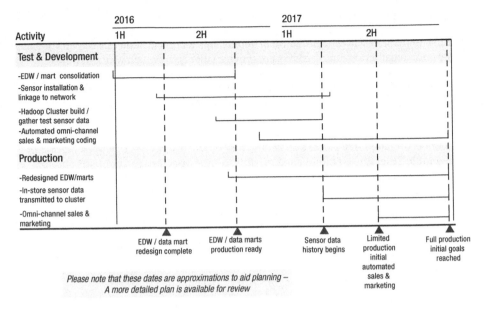

*Figure 7-3.* *Major project phases summarized for IT*

---

■ **Note**   As the technical presentation grows in detail and length, there is often the need for an executive level technical presentation for IT management that summarizes the delivery of key phases in the project. As with the detailed technical presentation, it tends to be less business-oriented than the roadmap presentation, which targets a broader executive audience that we will describe in the next section of this chapter.

---

Now that we've gathered together all of the information that we have discovered so far and we are convinced of the viability of the project, our next goal is to secure funding. We need to create a business executive level version of the content that will be relevant to decision makers in our organization and that will help sell the project.

# Building a Roadmap

A well-defined roadmap can help executives chart a path to a new destination. However, this roadmap must also provide the executives with reasons why they will want to follow that path, including why it will be worth the time and expense and how they can be assured that they will reach the destination. The roadmap generally takes the form of a presentation, though a summary document might also be prepared.

Remember that brief and to the point is the best approach when presenting to a broad executive audience. In the following flow, each major bullet point is represented by a slide in the example that we will illustrate. The content that should be covered is as follows:

- Our recommendation that we will also revisit at the conclusion of the presentation:
    - A one-minute summary of what business problems are being solved and the scope of the effort that tees up the fact that we will ask for funding at the close of the presentation
- Agenda:
    - The agenda describing what we will cover in support of our request for funding
- The business drivers:
    - A review of critical business drivers as provided by the lines of business, answering the question of why do anything
- Project phases and associated business case:
    - Project phase timeline and return on investment
- Technology solution:
    - Solution evolution from current state to future state
- Risks and mitigation:
    - A summary of significant risks to project success and steps to mitigate those risks
- Conclusion and next steps:
    - A summary that describes why the project must start now and a request for funding

Now, let's take a look at an example of a roadmap presentation. A roadmap presentation for executive audiences that describes our omni-channel project could be based on a set of slides that resemble the following figures. We begin with an upfront summary slide, as pictured in Figure 7-4.

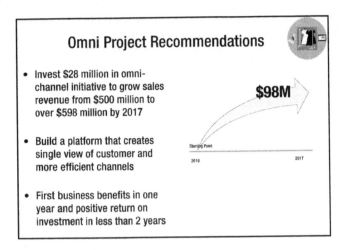

*Figure 7-4. Roadmap executive presentation—recommendations slide*

As you can see, we quickly will establish what the project will deliver (a single view of the customer and improved efficiency in our channels), the project's cost ($28 million), and highlight a measurable business benefit (the incremental sales revenue we expect to gain, $98 million). We will also highlight how quickly the benefits will be delivered (by the end of 2017).

---

■ **Note** We might achieve other business benefits in our project and, in fact, we will point to some later in our example presentation. However, on our initial recommendations slide, we want the focus to be only on benefits that we have solid and well-supported quantitative numbers for. It is much too early in our presentation to get into a confrontational discussion about our assumptions.

---

Of course, our executive audience will want us to back up these statements. As a result, our next slide, shown in Figure 7-5, illustrates the agenda that our presentation will follow in making our case.

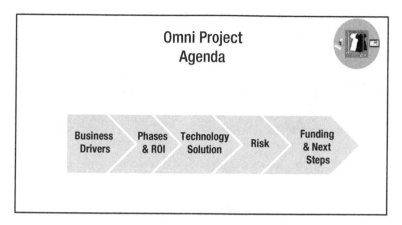

*Figure 7-5.* *Roadmap executive presentation—agenda slide*

Before moving on, we should ask our executives if they see anything missing from the agenda. They might mention particular concerns that we planned to cover anyway. However, this will help us determine where we might want to spend more time on the agenda item that is most appropriate.

Of course, our audience must understand the expected business results before they will make an investment and fund our project. In Figure 7-6, we provide an overview of the key business drivers for the omni-channel project. It should be clear that for this project to succeed as pictured by our stakeholders, it must provide a solution that can optimize promotions across the channels, optimize the supply chain across channels, and grow customer revenue across channels.

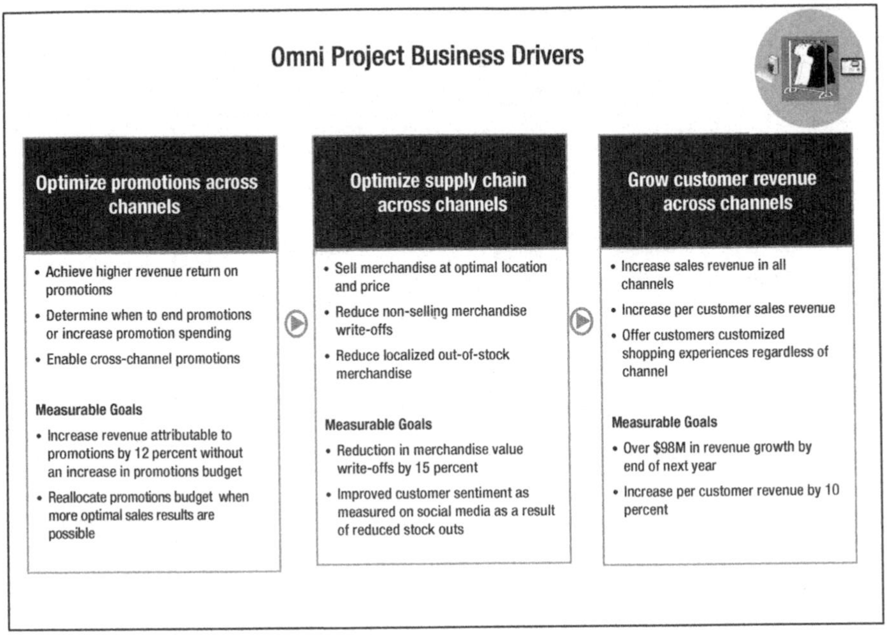

*Figure 7-6.* *Roadmap executive presentation—business drivers slide*

We also provide more details on desired outcomes and measurable goals on the slide. All of the measurable goals stated must be backed up as being reasonable by our key stakeholders (for example, well-respected leaders in marketing, supply chain management, and sales). These should be the same figures that we used earlier when we developed our business case. Our key stakeholders must be willing to stand behind these figures again and support the key goals of the project.

The executive audience will want to understand how the project might be broken into manageable phases that will begin to deliver the business return we just described. They will also want to understand if such phases can mitigate risk of project failure and cost overruns.

Figure 7-7 shows how we might illustrate the phases. The markers along the lower axis indicate the expected incremental revenue gained over time. A diagram like this is sometimes used to also illustrate the project costs over time as the benefits accrue.

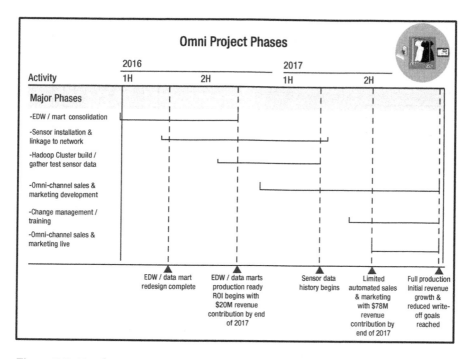

*Figure 7-7.* *Roadmap executive presentation—project phases*

The figure also denotes other activities that this audience might be concerned with (and you might compare Figure 7-7 and its activities to those shown in Figure 7-3, which was targeted toward IT management). For example, we indicate here that we will have a plan for change management and training that will be executed prior to the go-live phases. This was added to the diagram in anticipation of some of possible objections that the executives might raise around the ability of the lines of business to use the proposed solution as well as the ability of IT to manage it.

Despite laying out a solid business case, our audience might still wonder why the project is so expensive. It might not be obvious to them that we are describing major modifications to the infrastructure in the brick-and-mortar stores and within IT. These are being driven by the need to gather and analyze data across multiple channels in ways that don't exist today; we will need to show them.

Figure 7-8 represents our current state infrastructure in a simplified form. We can use this figure to point out that data about our customers is currently held in silos. As business leaders, they should already recognize that the organization can currently see what the customers bought in each of the channels, but that it has little ability to deliver on the three business objectives we just outlined, which require a cross-channel customer view.

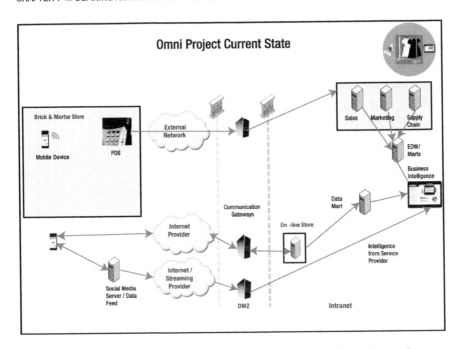

*Figure 7-8.* *Roadmap executive presentation—current state information architecture diagram*

Figure 7-9 is used to illustrate how the infrastructure must change. Sometimes, describing this change is most effectively accomplished by creating a build slide that illustrates a gradual transition to the future state and aligns to key project phases.

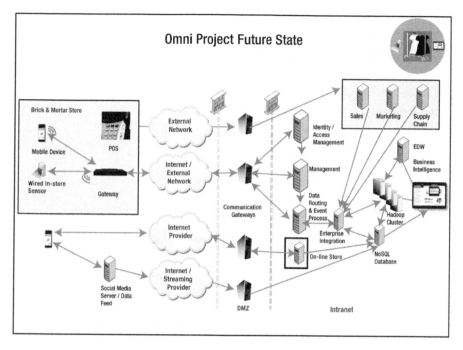

**Figure 7-9.** *Roadmap executive presentation—future state information architecture diagram*

We might anticipate that our audience could be taken aback by the complexity and other aspects of this project. Our next slide (Figure 7-10) illustrates that we have considered many of the risks and have a plan for risk mitigation. In addition to seeking their confidence in our plan, we might start to seek buy-in regarding training plans and the need for consultants as we discuss this. We could also use this discussion to open a dialog about all of the objections to the plan so far and get those out on the table.

## Omni Project Risk Mitigation

| Risk | Mitigation |
|------|------------|
| • Impact on business processes | • Continuous improvement & change management |
| • Competitive threat | • Agile architecture |
| • Design complexity & scope | • Incremental delivery, ROI |
| • Impact on analyst skills required | • Change management, training, consulting partners |
| • Impact on IT skills required | • Training & consulting |
| • Technology impact on store operations | • Remote IT support and on-site problem resolution |

*Figure 7-10. Roadmap executive presentation—identified risks and planned mitigation*

■ **Note** The executive presentation must establish and maintain credibility with the people who control project funding. We seek to confirm that we have a credible plan at every step along the way. If we fail to establish this trust, the project likely won't be approved. Consequently, enough detail must exist in the project plan behind this presentation so that we can confidently answer questions that will arise.

We should now be ready to close our presentation with a "next steps" discussion. We will ask for funding of the project and make note of anything that stands in the way. We also use the closing to describe any other next steps we have already anticipated, partly to show our confidence in gaining approval. Of course, we will make note of any other steps that the executives suggest and incorporate their ideas into our plans.

Our slide describing the next steps is illustrated in Figure 7-11. You might notice that we reiterated the anticipated incremental business benefit as well as the amount of funding requested on this slide. Continuing to highlight the potential return on investment drives home the message that while some might consider the project costly, it will deliver substantial value to the organization.

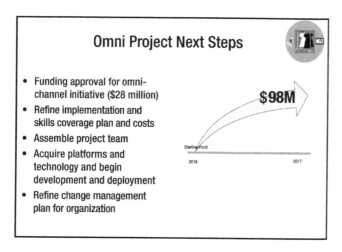

*Figure 7-11. Roadmap executive presentation—closing slide*

# Gaining Approval and the Transition

At this point, we should believe that our roadmap presentation is ready for executive review. We have done our homework, and the plan appears to be entirely viable to us. We also have our detailed plan well documented and can rely on the information in it when we respond to questions during the presentation. What could go wrong?

Unless the initiative came down as a CEO mandate, we will likely face a highly skeptical audience when we deliver the roadmap. The executives could be extremely risk averse and might believe that Big Data and the Internet of Things projects are simply overhyped, expensive initiatives that they want to avoid. Possibly, there is a bad track record in deploying innovative projects of this type within the organization. The executives could also have other hot buttons and business concerns that we are unaware of.

There are several steps we can take ahead of the executive meeting to be better prepared. First of all, we should review our content with our sponsors and seek their guidance on the materials that we have prepared. We might also review the content with some of the executives who will be present, or with their direct reports, in order to solicit their feedback (including any concerns they have or anticipate being raised). If we can meet with some of the decision makers ahead of the meeting, we might also seek their early endorsement for the project.

If we are told that our executive audience will need more visual proof to understand the value that our project will provide, we might build a demonstration showing some sample dashboards or build diagrams showing how business processes will be improved. We can review these demonstrations and other additions to the presentation with our "test" audiences before the executive presentation and further solicit input as to the value such visual aids will have in helping us make our case for funding the project.

# The Executive Meeting

Where our roadmap presentation appears in the executive meeting agenda and who is actually present during the presentation will indicate the importance of our project to the executives. Being placed last on the agenda with just a subset of the executives remaining is not a good sign. At least we are somewhat prepared if the presentation is cut short since, as described in the previous section, we have our recommendation, the business benefit, and what we are going to ask for summarized up front. In such situations, getting a full hearing at a later date might be the best we can hope for. We can use the initial meeting to start selling the value of the project in the limited time and to the limited audience we do have.

During the presentation, three things could happen. The best outcome would be approval for funding. However, we might be given the second outcome that no funds exist and that the project will not be considered. A third option is often the most popular in some organizations—deferring a decision on funding the project pending more information or the gathering of a different executive audience for the presentation. Let's take a look at what we might do if the latter two outcomes occur during the meeting.

If the project didn't receive approval because of a lack of funds, the real reason could be that the business case and / or the sponsorship are not strong enough. In many organizations, truly compelling business cases can drive a re-allocation of funds to a project. Since we have established a positive ROI, perhaps the problem is that payback occurs too slowly. We must simply ask what is standing in the way of a positive decision. A lack of funds might not be the real issue. It could be that the project is not out of the question, and we can come back later with answers to executive concerns and, thus, achieve approval for funding.

In some organizations, getting any sort of decision can be difficult. If the decision is deferred, we must find out exactly why the decision is being put off and ask how and when we can bring the project forward again in the future and assure that a decision will be made.

If the decision is made to establish funding and go ahead with the project, prior to leaving the meeting and celebrating, we should ask the executives whether they want regular reports about the progress of the project and how those should be delivered. We might suggest that five minutes be set aside during subsequent executive meetings to cover progress and bring up any issues encountered.

# Transitioning to Implementation

With funding in hand, now is the time to begin assembling the project team. The project sponsors might think their job is done once funding is obtained, but they should remain key members of the team throughout the implementation process. They will often serve as a key liaison into the lines of business and to their executive management. They can also reinforce how important the plan is to the business in meetings with the project team when there is a need to do so.

Of critical importance in putting together the project team is the naming of the project manager. The project manager will identify the resources needed, recruit or hire those resources for the team, set and negotiate milestones, and monitor project progress to make sure milestones are met within the funded budget. Project managers should be familiar with standard project management frameworks in place in the organization for tracking and monitoring of activities and tasks. They will also monitor the project for quality and changes in scope on an ongoing basis and bring together concerned parties for resolution when quality issues, scope creep, or issues among team members will impact the ability to deliver promised business objectives in a timely manner.

Once named, project managers will review all of the information gathered thus far in the detailed plan. They will question our early participants who helped formulate our initial plan if they don't understand why certain trade-offs were made. As they assemble their team, they will ask the team and their trusted advisors to poke holes in the plan as it develops, determine how the plan could fail, and then fix potential holes in the plan.

---

■ **Note** Great project managers are highly effective in their communications to a wide range of audiences, ranging from highly technical individuals and business experts to team leaders and corporate executives. They are well organized and are particularly effective at problem solving. They tend to be politically astute and know how to deliver bad news in ways that lead to innovative problem-solving strategies that move projects forward.

---

A variety of teams are commonly formed and become part of the broader project team. Some of these teams would likely be built around the following members:

- Business analysts and data scientists (including lines of business representation)

- Data management systems architects

- Data acquisition specialists

- Enterprise architects

- Infrastructure and design architects (servers, storage, and internal networking)

- IT operations personnel (for example, those familiar with organization's best management practices, standards, and service level agreements)

- Intelligent sensor development specialists and programmers (for Internet of Things projects)

- Networking design and operations personnel (for transmission of data from sensors to the analytics infrastructure)

Members within the teams are selected based on their ability to define and build solutions. Of course, they must also have interpersonal skills and the ability to work under the pressure of time and delivery scope deadlines. As we discussed during our skills evaluation process, the team members might come from internal employees, consultants, systems integrators, or a blended mix of some or all of these.

As members are added to the teams, the teams can grow so large as to require their own management chain, so team leaders are usually designated within each team. Team leaders are often sought with similar skills as the project managers given their critical role in the solution delivery process.

Skilled project managers also spend the period of time after funding approval and before implementation building more details into their project plans by using input they obtain from their teams. Figure 7-12 illustrates a simplified, high-level plan for ETL development that a data acquisition team might provide. In this format, such a diagram is sometimes referred to as a Gantt chart. Underneath this ETL plan would be another level of detail describing the building of solutions for extraction of key data from internal and third-party data sources and its loading into solutions supporting each business area (such as sales, marketing, and supply chain).

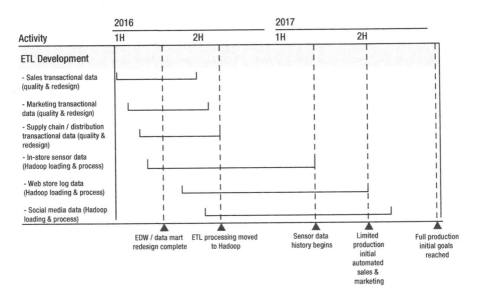

*Figures 7-12. Typical high-level ETL development project management plan*

As you might have guessed, a detailed overall project plan can consist of hundreds of activities and grow to contain thousands of tasks. Understanding how the activities and tasks are interrelated and how delays in specific tasks might affect key milestones is of particular interest to the project manager at this point in the planning. This is typically accomplished using the Critical Path Method (CPM), which identifies the tasks that are on the critical path to completion of the activity. A PERT (Performance Evaluation and Review Technique) chart is sometimes used as an alternative to the Gantt chart to represent task schedules since it clearly illustrates the tasks on the critical path.

Thus far, our work in these phases has consisted mostly of design and planning, so an architect typically will have had a leadership position. As you can see, we are transitioning project leadership to the project manager. Of course, the architect's job isn't complete yet, as you will see in our final chapter.

The last chapter describes the phase that will provide what our methodology is designed to produce: an implementation that enables the desired business solution. Similar to the other chapters in this book, we will focus on the best practices that we have observed during that phase of the project.

# CHAPTER 8

■ ■ ■

# Implementing the Plan

We conclude this book with a chapter that covers implementing the project plan. This is the phase in which all of our hard work in preparing the plan and obtaining funding should pay off. We will begin delivering the solutions that provide the value promised in our business case. Whether we are successful in this phase is due in part to how well we executed our methodology for success during earlier phases.

However, much still needs to be done before success can be claimed. Many pitfalls are possible during this final phase of the project. We will touch on many of them here. This phase can be particularly challenging given the innovative nature of projects that include Big Data and the Internet of Things technologies.

As we begin this phase, we have developed our project plan, named a project manager, and assembled our teams (as described in Chapter 7), and we are ready to begin creating and operationalizing the technology that will enable our new solutions. In this chapter, we start by describing a stepwise approach to implementation and related best practices. We also describe what might cause changes in the project's timeline that we should be on the lookout for.

Once developed, we must be ready to put our production-ready solutions into operation. We describe the transition that occurs when operationalizing our solutions from a technology standpoint and revisit some of the earlier design considerations that we described in Chapter 6. We also describe the importance of executing a change management plan during this phase.

When the project is fully deployed according to the defined scope, we will declare an end to the project. Many organizations have trouble with this step as these projects often evolve in response to requests for more capabilities from the lines of business. This is particularly true once the lines of business begin to visualize other possibilities that these solutions might provide. Of course, the success of our project must be established and reinforced in the minds of executives and key stakeholders. This will help us build trust in our ability to successfully deliver these projects.

At this point, we also need to re-evaluate how well our solution fulfills the original business case and look at any adjustments we made along the way. We should capture the lessons that we have learned so that they become part of our future best practices. We might also discover that new demands coming from the business are so substantial in scope that they could lead to the start of a new project with its own life cycle.

Our final section describes starting another project around new and earlier out-of-scope requirements. Though we will repeat our methodology for success, all participants involved in the next project will have learned a lot from the previous one. Business and technical expectations for success will likely be much higher.

Figure 8-1 highlights that we are at the project plan implementation phase in our methodology for success and illustrates that we divided this chapter into four sections: implementing the steps in the plan, operationalizing the solution, ending the project, and starting again. Though this is the final chapter in this book, we believe that a never-ending cycle can be established when projects are delivered that succeed in delivering promised value to the business.

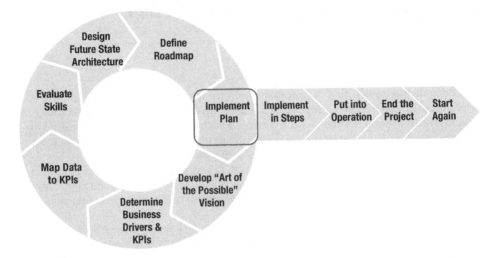

***Figure 8-1.*** *Implement plan phase in our methodology for success*

In the next section, we describe how we might go about implementing our project using a stepwise approach. As you will recall from earlier chapters, we do this in order to deliver incremental interim solutions that have a more manageable scope of effort while also demonstrating progress and business value along the way. Incremental solutions produced every 90 to 180 days are considered to have a reasonable delivery schedule since the time sequence is short enough to hold the attention of the business community.

# Implementation Steps

Once the project teams are assembled and our plan is in place, we are ready to seriously begin the project implementation phase. A common first step in the implementation phase is to hold a project launch meeting. This meeting should be much more than a formality. The launch meeting is an opportunity to get the entire team on the same page and establish common goals among the various participants.

The typical launch meeting should be an in-person gathering that is led by the management team. A web-based and telephone conference call meeting will not accomplish the goals of the launch. The meeting agenda usually includes the following:

- A discussion and overview of why the project is of significance to the business as provided by a very senior level business executive

- A brief history of how the business vision of the solution came to be reiterating its importance as described by the project sponsor

- An introduction of the team leaders and teams present

- A discussion of the planned evolution of the information architecture by the lead architect

- A description of the project plan, including a discussion of key milestones such as how scope, quality of effort, and progress will be monitored; how to raise concerns about the schedule; and what the planned schedule of project assessment meetings is as described by the project manager

- A description of how activities and tasks will be tracked and the importance of this tracking by an assistant project manager or senior team leader(s)

- Some final thoughts on the project provided by the project sponsor

A goal of the meeting is to get everyone excited about the project and enthusiastic about the teams that the implementers are part of. We also want team members to understand their role in bringing the project to a successful conclusion. Working relationships developed during this face-to-face meeting can result in better teamwork and save a lot of time later in the project when clean hand-offs and collaboration are crucial. This is also an opportunity for the individual teams to gather and discuss their role in the execution of their part of the plan.

When the meeting ends, everyone should have a clear understanding as to the activities and tasks that are their responsibility, the budget constraints in place, and the due dates for delivery. Teams should possess a common sense of purpose and understand specific performance goals, while the complementary skills of individuals on the teams should also be understood. All participants should have a strong commitment to the project, mutual trust in other team members, and an understanding that they will be held accountable.

Success of the project will be measured by the consistent delivery of well-scoped, incremental development efforts. Careful management of any changes must occur and will ideally remain within the project timelines that have been established.

## Project Plans and the Critical Path Method

In Chapter 7, we briefly introduced using a Critical Path Method (CPM) for identifying tasks that are on the critical path to the completion of activities and the project itself. Milestones are defined as the dates when tasks or activities must be completed. The project manager pays special attention to meeting these milestones as the project implementation is initiated and continues. Sometimes, a task not on the critical path will have an early start and early finish date that is defined as the first date the task might be started and completed. Also frequently declared is a late start and late finish date that is the last date that a task can be started and completed. The difference between the late dates and early dates is referred to as "float" or "slack" in the schedule.

Of course, tasks in the CPM network of activities are dependent on earlier tasks. As a result, ongoing tracking is a must to understand how well the project is progressing. Most teams will also use CPM to track the ongoing costs of tasks and activities as the project proceeds in order to detect and mitigate possible cost overruns that could occur.

---

■ **Note**    Some project managers prefer to have variations in finish dates and lengths of tasks and activities in their CPM models to account for the natural tendencies of some individuals to be conservative when providing time estimates, while others are not conservative enough when estimating the amount of time needed to finish a task. Time estimates also tend to be shorter when the project is being sold and become longer when the full scope of the task at hand is realized.

---

To illustrate how CPM works, we will refer to Figure 8-2, which shows a small portion of a critical path diagram in a development effort. Parallel tasks are pictured that start on the same date (3 June), but one of the paths becomes the critical path. Node identifiers in the model are indicated by the four digit numbers within the circles in the diagram. The tasks begin at an originating node and continue for a duration shown over the connection between nodes. For simplification regarding task completion dates, we are assuming that the team is working seven days a week.

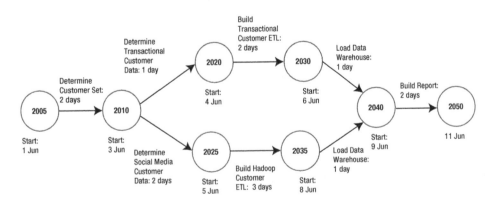

*Figure 8-2.* *Simple critical path method diagram*

The critical path in this diagram is the one that follows building the ETL code for the social media customer data since it takes longer to execute this series of tasks. Hence, the start date for building the report from data in the data warehouse is 9 June, even though the transactional customer data could populate the data warehouse as early as 7 June.

Given that the building of the ETL code for the customer transactional data is not on the critical path, we could assign a late start date for this task to be 6 June (instead of the pictured 4 June). The late finish date for the task of building this ETL would be 8 June and so we would then load the data warehouse with this data on the same day as we load the social media data. This would not delay the start date of building the report.

The project manager is on the lookout for tasks and activities that are taking unusually long and / or requiring more resources than originally planned. These inconsistencies could be due to a variety of reasons including original time and resource requirements that were badly estimated, lack of properly skilled individuals to implement the task, or scope creep caused by new demands and changes. Any of these could lead to missed milestones and a need to re-analyze the validity of the plan. Personnel choices are sometimes under review during particularly challenging moments.

To help assure that milestones will be met, project managers will sometimes create "management reserve" tasks on the critical path prior to work getting underway. These tasks are not actual work, but are simply additional time buffers. The percent of additional time that is added is usually dependent on how well the project manager and team leaders know their teams' ability to deliver the needed tasks. Adding 10 to 15% of additional time is typical. The management reserve tasks are invoked when the time to implement tasks overruns the allotted time on the critical path.

# Best Practices for Driving Timely Progress

Although some participants might desire project update meetings to only be held as milestones are reached, this practice can lead to a backlog of problems that introduce delays that are difficult to recover from. So, most project managers prefer holding weekly update meetings with the various teams. Of course, if major problems are uncovered in the interim, the project managers want to be notified immediately.

One technique that is sometimes used during a critical development phase is to develop the solution far enough along to become a working prototype that can be demonstrated to key business sponsors and analysts. In Big Data and Internet of Things projects, a prototype can prove to be extremely valuable in providing an early test of the integration of components as well as the ability of the solution to deliver what the business is expecting. This approach can help to assure that when the milestone is reached from a technical development standpoint, the lines of business will agree that the solution delivers what was promised.

Quality of the deliverables must also be tracked. The project manager must understand whether the deliverable meets the specifications for the task at hand. Failure to monitor quality could have negative implications as substantial rework might be necessary at the time an incremental solution is believed to be nearly ready for delivery. Quality can be assessed through a variety of techniques including peer review, outside consultant assessment, and review by line of business early adopters.

As an incremental solution nears completion, resilience and functionality testing of the solution is necessary. This testing occurs to make sure that finished development effort can handle input data volumes and meet other requirements while delivering the scoped solution under all conditions. Any exceptional situations should be accounted for in the design and implementation.

Project progress is reported on a regular basis to key stakeholders at review sessions. Within the project team, management reviews occur on an ongoing basis to better manage risk, especially where time and cost estimates are in question or the scope of the task or activity is unclear. Regular reviews are also held as milestones are reached and where some of the expected business value should be delivered. At such critical junctures, decisions are sometimes made as to whether the project should go forward before further changes are made, and technical reviews might be undertaken at this time.

---

■ **Note**    Sometimes the costs already sunk in a project establish a momentum to continue onward with development, even if the business case for the project becomes questionable. At such times, it is necessary to separate the emotion associated with the investment made thus far from a re-evaluation of the business case to determine if proceeding with the project in its current form makes sense.

---

Changes in the project timeline are never desirable. However, in complicated projects such as those typical in Big Data and Internet of Things implementations, they are sometimes unavoidable. Whenever they occur, proper communication of the status of the project to all concerned is a must. Changes in the timeline and the impact on the delivery of business benefits and cost of the project should be of wide interest. Important decisions such as changing the order in which incremental solutions are delivered might be up for discussion. Priority changes need to be discussed with business stakeholders as well as technical implementation teams.

Any changes in project scope should be handled through a formal change control process. Implementing a formal process helps the project manager control changes to the project scope. A formal project change request should be submitted to the project manager for review and approval or rejection as is appropriate. When such changes are proposed, the project manager must be given enough information to understand not only the scope of the change, but also the impact on the schedule, cost, quality of the deliverable, and the resources that will be required. The risk to successful and timely delivery of incremental solutions and the overall project solution must be understood.

## Causes of Change to a Project Timeline

There can be many causes of unavoidable change to a project timeline. For Big Data and Internet of Things projects, there can be skills issues and complicated technology implementation and management challenges. However, there can also be adjustments to the project and its timeline that are not caused by the technology, but are instead caused by changes in business requirements.

Given the demand for talented individuals in projects of this type, personnel skills management issues are frequently faced. Skilled individuals and consultants are often difficult to find. Team members sometimes receive attractive offers elsewhere during the course of the project and replacements can be difficult to secure when team members decide to leave. Monitoring the management of valued personnel and establishing a bench of talent in reserve is of great importance.

Where partners outside of the organization provide systems development and integration or engineering skills, solution development efforts can be slowed by the inability of the partners to create deliverables that match the scope and quality requirements. The resumes shared during partner evaluation might not align with the skills of the people staffing the project. Some changes in the staffing provided might need to be negotiated.

As the implementation proceeds, it can become evident that while developers on the teams have the skills required for prototype development, they might not have the skills needed to create solutions ready for deployment and management required in an enterprise infrastructure. This can become especially apparent when operationalizing a solution in an organization unfamiliar with Big Data and / or Internet of Things technology components. Additional resources that understand the enterprise's operational standards should become part of the team if this skills gap is found.

Development efforts might also be challenged by rapidly changing key technology elements that are part of the solution. For example, major Hadoop releases and NoSQL database releases can occur several times in a year. Each release can include significant new features that solve old technical problems in better ways, but sometimes require new coding efforts. Decisions must be made as to whether to recode a solution to take advantage of the new features or to continue the development path that was already underway.

Many other challenges to meeting milestones are possible.

In Internet of Things projects, a lack of availability of the right sensors in proper device locations and a resulting inability to gather needed data can lead to project delays. It can become necessary to work in close partnership with a device manufacturer to resolve this type of problem or instead to engineer a custom solution of our own. Close coordination between engineering teams and software development teams is required.

Communications backbone providers can also have a critically important role in Internet of Things projects. Their designs and implementation will require close collaboration with the engineering and software development teams. Communications backbone providers that lack experience in these types of solutions can throw roadblocks into the project implementation as network bandwidth, security, availability, or other issues arise.

On the business side, challenges to the success of the project can occur due to organizational changes or because of the departures of key business sponsors. Business analysts who are driving the use of the more advanced solutions that are being built might transfer to different roles in the organization and could be replaced by less sophisticated analysts who are not able to use advanced solutions. The business needs could also change due to new competitive pressures facing the organization or due to a change in business direction as directed by senior executive leadership.

---

■ **Note** Any of these changes introduce new risk to the success of the project plan. Close coordination among all involved is critical to managing risk and making changes to the plan. Of course, concerns can and should be raised. If significant concerns are not of interest to project sponsors and executives, perhaps the project is not considered all that important. That, in itself, could be a sign that there are problems ahead.

---

# Operationalizing the Solution

As milestones are reached and project phases are ready to come online, basic development ends and operationalizing of the solutions will begin. It is at this point that the solutions must support agreed-upon service levels for successful delivery. A number of the availability, management, and data governance considerations that were introduced as design considerations are re-evaluated at this time. We also document our operational procedures for IT and review how well we have documented our ETL processes for the business analyst community.

We must also take another look at the skills enablement needed for IT to manage these solutions and for business analysts to use the solutions effectively. During this stage, we put into practice change management techniques that require close collaboration and training. We also begin to sell the business capabilities that are being delivered to a wider audience in order to generate further enthusiasm for the project.

## Service Levels and Documentation

As we gathered project requirements earlier, we began to understand needed service levels. If not already well defined, we must define service level agreements (SLAs) in detail here. The SLAs will drive our deployment strategy and configuration to match performance, reliability, availability, serviceability, and security needs.

For example, we should have a clear picture of how many business analysts will use our solution when it reaches production and understand the workload that is likely. We also should have an idea as to how large the analyst community could eventually become and their future workload demands. Prior to releasing the solution into production, we will want to test the solution to scale so that we become confident that we can meet or exceed the SLAs that are agreed to.

We will also test management best practices required to meet SLAs prior to going into production. Among the procedures that should be tested are the following:

- Day-to-day operational management capabilities including our ability to maintain data and solution availability in the event of failure in the software, server, or communications infrastructure or in the event of a disaster

- Monitoring of the entire infrastructure including event management

- Problem and incident management and patching

- Performance management and tuning

- Infrastructure change and application release management

- Business user management and data protection management including access controls, data encryption, and auditing

During this testing, we will develop the operational procedures that must be put in place and document the procedures. We should establish clear support procedures within our operations team and through our server, software, and communications vendor relationships.

For example, we might define a set of procedures for updating applications hosted on a Hadoop cluster. As illustrated in Figure 8-3, the application release management process would begin with initial testing of the new version of the application in a development environment that is quite different from the production environment. Such early testing would take place in a "pseudo distributed environment" consisting of virtual machines. The pseudo environment is used to validate that the application is ready for more advanced testing in an environment that is similar to the production environment. The advanced testing enables us to determine the ability of the new version to meet SLAs for functionality, performance, and availability. After testing succeeds there, the application might next be deployed to the cluster disaster recovery site and then to the primary production site.

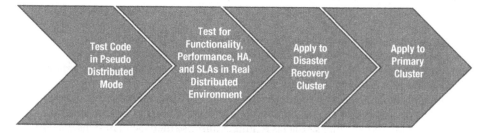

**Figure 8-3.** *Application update procedures on a Hadoop cluster*

Part of the application release management documentation might also include references to proven backup and recovery processes for the cluster, just in case something goes wrong during the process. We could define how to use Linux operating system utilities to create backups of the program files, source code, and configuration files that reside on the cluster. As the data in Hadoop clusters is of such volume that recovery time objectives would be impossible to meet with classic backup and restore procedures, we would describe where data is duplicated and how to manage test data sets during application testing.

The step-by-step documentation we write describing this approach and others will establish repeatable processes that can be successfully put into practice by the operations staff. Those who struggle with the documented steps and process during testing might be candidates for further training (or the documentation might need further improvement).

---

■ **Note**    When we develop our solutions, we should define technical and business metadata during ETL coding and through the business intelligence tools that are deployed. This is another important part of the documentation process as it enables the business community to understand what the data represents during data lineage analysis and when using analyst tools.

---

## Organizational Change Management

Experienced project managers recognize the importance of focusing on change management at this point in operationalizing the solution. If the business analysts are not prepared to use the solution as it is delivered, the project is almost certain to fail and further funding could be dropped. Positive return on investment for the project is usually dependent on the business community understanding how to derive the maximum value possible from the solution and using it to deliver measurable business results.

Change management procedures that are created for the lines of business often include a training plan that will enable analysts to gain the skills needed to use the new analysis tools that are being deployed. Generic tools training usually provides limited value to the analysts. A more appropriate training plan will focus on how to obtain the business intelligence from the data in a way that matches the business goals for the project.

This is also often the point at which ownership of the data itself might be transitioned to the lines of business. If data ownership is transitioned during the operationalizing of the solution, a training plan for the new data stewards, data sponsors, and data custodians should also be developed.

Similarly, a training plan is often developed for the IT organization in anticipation of new systems and data management solutions coming online. When in production, Big Data and Internet of Things projects require enterprise management skills that are more akin to traditional skills provided for data warehouses and other IT infrastructure. As we've discussed elsewhere, such skills are usually in short supply and / or need to be reconsidered when the projects move beyond the research and development phase. As a result, additional training is likely to be required.

Early in the deployment of our incremental solution, we might observe that the business value we anticipated is not fully realized. We might also see that management of the infrastructure is a problem. This should give us pause and cause us to re-evaluate the project timeline and future phases and readjust our change management procedures. We should get at the root of the problem before we go too far forward in a direction that could prove to be unsuccessful.

> ■ **Note**  When business value can be clearly demonstrated after an incremental solution has been deployed, we should begin advertising success. We do this to maintain project momentum and excitement within our development teams, but also to encourage more widespread business adoption. Of course, we need to avoid having excited business converts demand too many new requirements that cause scope creep beyond our capabilities to deliver according to the project timeline. However, we can start gathering new requirements for future stages or future projects at this time. This is a great problem to have.

# Ending the Project

One of the challenges that we have observed in many organizations is ending a project of this type. The demand for enhancements frequently stretches the time and scope of the original project. But, at some point, the project must end so that it can be fully integrated into the organization.

The end to the project occurs when all critical tasks are complete and a mutually agreed-upon final solution is delivered. A discussion and determination of transition dates where there is a transfer of project ownership, completion of remaining training, and start of any extended support and warranty coverage (if deliverables are under a warranty by an outside party) should occur. A formal acceptance should be recorded when all of the remaining in-scope tasks are completed to the organization's satisfaction.

When the project ends, it is critically important to claim success even as other projects might already be teed up. The old adage that everyone backs a winner is especially true here. In addition, a postmortem analysis of the project is in order to learn what might be done better when the next project is planned and executed.

## Claiming Success

If we have followed our methodology during this phase, we should now have a list of ways the project delivered business value during our incremental solutions rollout. We gathered objective measurable monetary values and subjective opinions as to the benefits provided. We also have the real incremental costs of delivering these project phases and of the overall project. Now we must take another look at our original predictions in our business case and roadmap and compare the earlier predicted business value to what actually occurred.

It is likely that we will find that some incremental solutions delivered far more value than we anticipated while others delivered far less. This is normal. What is important is that when we add up the measurable monetary value to the business, the benefits obtained from the entire solution exceed the cost of the project. We must make certain that the lines of business executives and sponsors recognize this and will back our conclusion.

We should also explore how benefits might continue to accrue over time. Of course, we should include in the future ROI the ongoing costs of maintaining the infrastructure. Some of the incremental solutions might not yet have paid for themselves early in their deployment, but they might produce positive ROI given more time. It is also possible that they were prerequisites for other more profitable phases that occur later in the project.

# Postmortem Analysis

Before we look forward to a new project, we should look back at the one we are completing and gather the lessons that we have learned. This will enable us to be much smarter the next time when we are developing a business case, architecture, and project plan, and when we are managing the project.

Smart project managers keep an ongoing list of reminders of what went wrong during various phases of a project. They measure individual and team performance and the impact of changes along the way. As the project nears completion, they often will interview team leaders and key team members, seeking opinions on what worked really well and what didn't during the project execution. A reassessment of the quality of the deliverables and the impact of quality issues on the project's success might occur. Project managers and other team leaders might also interview key individuals in the lines of business regarding their views as to how well the project proceeded and how it is delivering on its promises.

Some of the questions that might be asked during the postmortem analysis include the following:

- Did solution deliverables match business expectations? If not, why not?

- Were project milestones met on time and within budget? If not, why not?

- Did deliverables move smoothly from development into production? What, if any, issues became apparent?

- Was the original budget for the overall project accurate? If not, why not?

- Which teams were the most efficient and effective? Why?

- Which teams were the least efficient and effective? Why?

- What skills were found wanting, developed along the way, or are still in short supply?

- Did quality assessment practices improve the project's deliverables? How?

- Is business adoption of the project's solutions delivering the predicted business value?

- Were technical operations and business change management programs adequately planned for and well executed?

---

▪ **Note** Architects who provided leadership roles in earlier phases of the methodology should also take part in the interviews during this phase. The postmortem analysis provides an opportunity to understand where the architecture succeeded and where it led to implementation problems. The architecture will also be evaluated for how well it delivered on the promised goals of the project. There are many lessons that an architect can learn and apply to subsequent designs after a project is complete and in production.

---

Once this information is gathered, the project's outcome should be reviewed with key executives, sponsors, and other leaders in the business analyst community. They will also likely have views about what worked well and the challenges ahead and might not be fully aware of the project's benefits. The project manager should be prepared to present the following:

- A summary of business results attributable to the project and early feedback

- A timeline of how the project progressed and came online

- Challenges that occurred during project development and rollout and how they were overcome

- Lessons learned

- Deferred modification requests during development of the project and next steps

The gathering of all of this information can lead to a fair assessment of the project's outcome. A detailed report should be prepared documenting the project's goals, challenges, lessons learned, and the results achieved. It should be widely shared among the teams and with key sponsors. This will help demonstrate that a very thorough and consistent approach was used throughout the entire methodology. It can further establish the credibility of the project leadership and the work of the teams, and it will prove extremely useful when selling the next project in order to gain funding approval.

Of course, the final step in a successful project is often a gathering held in celebration. This is an opportunity to recognize and acknowledge everyone's effort in making the project a success—from project team members, team leaders, and architects to key sponsors, executives, and business analysts who provided support and guidance. One of the goals is often to reinforce a feeling of accomplishment and to encourage outstanding team members to consider working on envisioned future projects. Some managers also use this opportunity to begin discussing new projects with their potential future sponsors while the glow of success from the previous effort is apparent.

# Starting Again

Our methodology for success is complete. We have traversed the entire methodology, delivered our project, and reviewed the outcome. Now that we have progressed through the entire methodology, when we prepare to move onto a new project, we will not be starting from scratch.

We have now established a track record that we can point to. Hopefully, that track record is a good one. If the next project is an extension of our previous one, chances are that we have already gathered a lot of the requirements. Regardless, we should repeat the methodology for success as we begin anew.

---

■ **Note** In some organizations, the project team that developed a project will also bring it into production and then become fully consumed by ongoing management of the finished solution and providing further enhancements. This approach can severely limit the ability of the organization to take on new projects. A more desirable approach is to pass control of a completed project to an operations team. This enables the project team to begin to focus on developing a new project, and it enables the organization to take advantage of the many project management and development lessons that were learned previously by this team.

---

We will start once again by holding a discussion with the lines of business and IT about the "art of the possible" and jointly lay out a vision with them. If our previous project was successful and many of the same individuals are present in the planning of the new project, we should have a much more interactive and enlightened discussion than would have been possible the first time through the process. Much of what we gathered previously from the lines of business and our understanding of the technology footprint could prove useful as we begin establishing a new vision.

As before, we will next gather more detail regarding potential business drivers, critical success factors, and the measures and key performance indicators required, and we will begin to establish a potential business case. We will next map the required data sources to the output needed and describe the processing of the data that takes place. Then we will define the information architecture to be deployed in more detail. A roadmap to implementation will be created to enable us to sell the project and gain funding. Once funded, we will build a project plan in more detail and begin another project implementation.

Figure 8-4 illustrates the now familiar figure of how we progress through our methodology for success. You should now also begin to understand why we represent it as a closed circle. As the figure illustrates, we understand what is ahead but also draw upon our previous experience as we begin each new project.

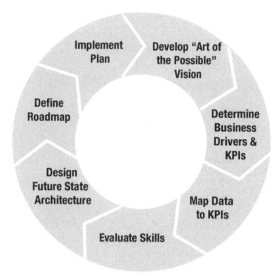

**Figure 8-4.** *The methodology for success as a closed circle*

As these cycles are repeated, the analysis of the current state, design of the future state, and odds of success will continue to improve by using this repetitive approach. We will gain a much clearer picture as to how well various teams and individuals perform when developing projects of this type, and we will increasingly know where to look for problems. We will also better understand the technology limitations and how to overcome them.

These projects should become much less riskier to business sponsors and to the technical teams by building upon previous successes. That, in turn, should lead to more funded projects and opportunities for everyone—the architects, project managers, developers, business analysts, business executives, and anyone else connected to solution development and utilization of these solutions in the organization.

# APPENDIX A

# References

## Published Sources

Arraj, Valerie. *ITIL: the basics* (white paper). The APM Group and The Stationery Office, 2013.

Buytendijk, Frank. *Performance Leadership*. New York, NY: McGraw-Hill, 2009.

Chodorow, Kristina. *MongoDB: The Definitive Guide*. Sebastopol, CA: O'Reilly Media, 2013.

Farooq, M.U., M. Waseem, A. Khairi, and S. Mazhar. "A Critical Analysis on the Security Concerns of Internet of Things (IoT)." *International Journal of Computer Applications*, February 2015.

Fields, Edward. *The Essentials of Finance and Accounting for Nonfinancial Managers*. New York, NY: AMACOM, 2011.

Greenwald, R., R. Stackowiak, and J. Stern. *Oracle Essentials: Oracle Database 12c*. Sebastopol, CA: O'Reilly Media, 2013.

Koster, Michael. "Data Models for Internet of Things." *Open Mobile Alliance*, February 2015.

Harvard Business Review. *HBR Guide to Project Management*. Boston, MA: Harvard Business Review Press, 2012.

IEEE-SA. "Internet of Things Ecosystem Study." New York, NY: IEEE, 2014.

Inmon, W.H. *Building the Data Warehouse*. Indianapolis, IN: Wiley Publishing, Inc., 2005.

Ishaq, I., D. Carels, G. Teklemariam, J. Hoebeke, F. Van den Abeele, E. De Poorter, I. Moerman, and P. Demeester. "IETF Standardization in the Field of the Internet of Things (IoT): A Survey." *Journal of Sensor and Actuator Networks*, 2013.

Josey, Andrew. *TOGAF Version 9.1 Enterprise Edition, An Introduction* (white paper). San Francisco, CA: The Open Group, 2011.

Khosla, Sanjay, and Mohanbir Sawhney. *Fewer, Bigger, Bolder*. New York, NY: Portfolio / Penguin, 2014.

Kimball, Ralph, and Margy Ross. *The Data Warehouse Lifecycle Toolkit*. New York, NY: John Wiley & Sons, 2013.

Linoff, Gordon, and Michael Berry. *Data Mining Techniques*. New York, NY: John Wiley & Sons, 2011.

Miller, Simon, and William Hutchinson. *Oracle Business Intelligence Applications*. New York, NY: McGraw-Hill Oracle Press, 2013.

Mosley, Mark. *DAMA-DMBOK Functional Framework* (white paper). DAMA International, 2008.

Phillips, Joseph. *IT Project Management: On Track from Start to Finish*. New York, NY: McGraw-Hill, 2010.

Plunkett, T., B. Macdonald, B. Nelson, et al. *Oracle Big Data Handbook*. New York, NY: McGraw-Hill Oracle Press, 2013.

Ross, J., P. Weill, and D. Robertson. *Enterprise Architecture as a Strategy: Creating a Foundation for Business Execution*. Boston, MA: Harvard Business School Press, 2006.

Schrader, M. D. Vlamis, M. Nader, et al. *Oracle Essbase and Oracle OLAP: The Guide to Oracle's Multidimensional Solution*. New York, NY: McGraw-Hill Oracle Press, 2010.

Stackowiak, R., J. Rayman, and R. Greenwald. *Oracle Data Warehousing and Business Intelligence Solutions*. Indanapolis, IN: Wiley Publishing, Inc., 2007.

The Open Group. "The Open Group Certified Architect (Open CA) Program." Reading, UK: The Open Group, 2013.

Wheelan, Charles. *Naked Statistics: Stripping the Dread from the Data*. New York, NY: W.W. Norton & Company, 2013.

White, Tom. *Hadoop: The Definitive Guide*. Sebastopol, CA: O'Reilly Media, 2011.

Yinbiao, Dr. Shu et al. *Internet of Things: Wireless Sensor Networks* (white paper). IEC, 2014.

# Web Site Sources

The Apache Software Foundation. http://www.apache.org.

CEB, Business Outcomes from Big Data, 2013.

http://docs.media.bitpipe.com/io_10x/io_102267/item_674000/Business%20Outcomes%20from%20Big%20Data.pdf.

Cloudera. http://www.cloudera.com.

Deloitte, Enterprise Value Map, 2004. http://public.deloitte.com/media/0268/enterprise_value_map_2_0.pdf.

EAdirections. http://www.eadirections.com.

Hortonworks. http://www.hortonworks.com.

ITIL (Axelos). http://www.axelos.com/itil.

MongoDB. `http://www.mongodb.com`.

Oracle Enterprise Architecture. `http://www.oracle.com/goto/EA`.

San Diego State University, Data Flow Diagram Tutorial. `http://www-rohan.sdsu.edu/~haytham/dfd_tutorial.htm`.

The Open Group, TOGAF. `http://www.opengroup.org/subjectareas/enterprise/togaf`.

See also the web sites of standards bodies, open source projects, and consortia listed in Appendix B.

# APPENDIX B

▪ ▪ ▪

# Internet of Things Standards

In Chapters 1 and 6, we mentioned that a variety of standards, open source projects, and consortia were present in influencing the direction of Internet of Things (IoT) components at the time this book was first published. In this appendix, we list some of the entities that you are likely to encounter as you evaluate options in deploying IoT projects. This list is not complete, and we expect that it will continue to grow rapidly and change over time.

## Standards Bodies

### IEC - International Electrotechnical Commission (http://www.iec.ch)

The IEC has produced numerous IoT white papers including a paper on Wireless Sensor Networks (WSNs), and it has taken part in key standards development such as IEC 62056 (DLMS/COSEM) for smart meters and OPC-UA for data exchange among applications. See also the work of the ISO/IEC joint technical committees below.

### IEEE - Institute of Electrical and Electronics Engineers (http://www.ieee.org)

The IEEE Standards Association (IEEE-SA) has produced an IoT Ecosystem Study and is developing the IEEE P2413 standard for an Internet of Things architectural framework. At the time this book was first published, P2413 was in early draft.

Other IEEE standards are also frequently cited in building out IoT infrastructure. For example, IEEE 802.15.4 defines a standard for low-data-rate, low-power, short-range radio frequency transmissions for wireless personal area networks (WPANs).

### IETF - Internet Engineering Task Force (http://www.ietf.org)

IETF has focused on network routing protocols (for example, IPv6 packets) and is working on standards for Constrained RESTful Environments in the IETF CoRE working group. These efforts are addressing how to deploy self-organizing sensor networks interconnected with IPv6 and building applications using embedded web service technology.

### ISA - International Society of Automation (http://www.isa.org)

ANSI / ISA-100.11a-2011 for "Wireless Systems for Industrial Automation: Process Control and Related Applications" was approved in September 2014 and published as IEC 62734. It provides a definition of reliable and secure wireless operations including monitoring, alerting, supervisory control, open-loop control, and closed-loop applications. After initial approval by ANSI in 2011, compliant device production began in earnest and over 130,000 connected devices had appeared by the end of 2012. ISA / IEC 62443 (formerly ISA-99) provides a standard for automation system security.

### ISO - International Organization for Standardization (http://www.iso.org)

ISO standards relevant to IoT include ISO18185 for RFID and numerous other supply chain and sensor standards (ranging from device interfaces designed to monitor conditions to sensor networking and network security frameworks). At the time of publication, ISO/AWI 18575 was planned to address products and product packages for IoT in the supply chain. ISO is often seen as providing a valuable resource for reference architectures, specifications, and testing procedures.

### ISO/IEC JTC/SWG 5

This joint technical committee (JTC) of ISO and IEC produced a subcommittee / working group (SWG) that identifies market requirements and standardization gaps. It documents standardization activity for IoT from groups internal and external to ISO and IEC. Areas of collaboration this SWG focuses on include accessibility, user interfaces, software and systems engineering, IT education, IT sustainability, sensor networking, automatic identification and data capture, geospatial information, shipping, packaging, and thermal performance and energy usage.

### W3C - Worldwide Web Consortium (http://www.w3.org)

In early 2015, W3C launched a Web of Things initiative to develop web standards based on IoT and what it calls "a web of data." Many previous W3C standards efforts are fundamental to IoT development including XML, SOAP, WSDL, and REST.

# Open Source Projects

Open source projects are based on the notion of a shared code base with multiple committers or contributors. Within IoT, a number of such projects have emerged. Though not considered as standards in the classic sense, these efforts can become defacto standards if widespread adoption occurs.

### Allseen Alliance (http://www.allseenalliance.org)

This alliance of over 140 members (as of early 2015) created "AllJoyn," an open source framework used in developing IoT projects. The alliance is largely made up of non-IT companies interested in building IoT solutions. The framework that was created defines data and power transports, language bindings, platforms, and security methods, as well as providing a growing array of common services and interfaces.

### Contiki (http://www.contiki-os.org)

Contiki provides an open source development environment (written in C) used to connect low-cost and low-power micro-controllers to the Internet (IPv6 and IPv4). The environment includes simulators and regression tests.

### Eclipse (http://iot.eclipse.org)

Eclipse provides frameworks for developing IoT gateways including Kura (Java and OSGi services) and Mihini (written in Lua scripts). Industry services are bundled in a SmartHome project consisting of OSGi bundles and an Eclipse SCADA offering. Tools and libraries are provided for Message Queuing Telemetry Transport (MQTT), the Constrained Application Protocol (CoAP), and OMA-DM and OMA LWM2M device management protocols.

### openHAB (http://www.openhab.org)

An open source project called openHAB produced software capable of integrating home automation systems and technologies through a common interface. It can be deployed to any intelligent device in the home that can run a Java Virtual Machine (JVM). It includes a rules engine enabled through user control and provides interfaces via popular mobile devices (Android, iOS) or via the web.

### ThingsSpeak (http://www.thingspeak.org)

ThingSpeak provides APIs for "channels" enabling applications to store and retrieve data and for "charts" providing visualization. ThingHTTP enables a device to connect to a web service using HTTP over a network or the Internet. Links into Twitter are also provided for notifications.

# Consortia

Consortia are usually alliances of convenience where vendors and developers team to solve specific problems and share best practices. One of the goals is often to establish critical mass behind an emerging standard and develop support for formal adoption of the standard.

Within IoT, most initial consortia addressed broad horizontal IoT architecture challenges. Later, other organizations became involved that are aligned around specific industries and that define the IoT solutions required for those industries.

## ADA - Application Developers Alliance (http://www.appdevelopersalliance.org)

The Application Developers Alliance serves as an advocate on behalf of developers. It shares innovative strategies, such as those in IoT initiatives, through its publications. The Emerging Technologies Working Group provides information on IoT projects in the automotive industry, manufacturing, and retail, and it has also investigated the impact of wearable devices and IoT in the home.

## Continua (http://www.continuaalliance.org)

Continua is an alliance of more than 200 companies around the world focused on establishing a system of inter-operable personal healthcare solutions. The Personal Connected Health Alliance (PCHA) falls under HIMSS and is intended to represent consumers. Continua has developed a certification process for connected healthcare devices to assure they are easy to use, less labor intensive, free of inefficient technology duplication, and not prematurely obsolete.

## HGI - Home Gateway Initiative (http://www.homegatewayinitiative.org)

HGI consists of broadband service providers and vendors of digital equipment for the home. The HGI Open Platform 2.0 suite gathers home gateway software modularity requirements and provides remote testing tools.

## IPSO Alliance (http://www.ipso-alliance.org)

The IPSO Alliance provides a resource center and seeks to provide thought leadership in establishing the Internet Protocol (IP) as a basis for connecting "Smart Objects" in IoT. Their Smart Objects feature common design patterns and use the IETF CoAP protocol between the devices running Smart Objects and connected applications.

## IPv6 Forum (http://www.ipv6forum.com)

The IPv6 Forum is an international society consisting of regional and national chapters from around the world. This society is focused on promoting IPv6 usage, including in IoT projects, and supplies documents and news regarding IPv6 on its web site. It also produces Forum events on various topics including IoT.

## ITU - International Telecommunications Union (http://www.itu.int)

ITU-T provides recommendations that act as defining elements in information and communications technologies (ICTs). Their Global Standards Initiative on the Internet of Things (IoT-GSI) has been defined in Recommendation ITU-T Y2060.

## OASIS - Organization for the Advancement of Structured Information Standards (http://www.oasis-open.org)

OASIS seeks to drive the convergence and adoption of open standards for global information. In 2013 it adopted MQTT as its standard messaging protocol for IoT.

## OGC - Open Global Consortium (http://www.opengeospatial.org)

The OGC is an international consortium of over 500 companies, government agencies, and universities banded together to create geospatial interface standards thorough consensus. The OGC's Sensor Web Enablement (SWE) standards focus on gathering location data from sensors and address areas such as developing models and XML code for observations and measurement, the planning of data collection, and providing a common data exchange model.

## OIC - Open Interconnect Consortium (http://openinterconnect.org)

OIC provides a connectivity framework that enables common discovery and connectivity tasks across a variety of transports such as WiFi, Bluetooth, Zigbee, and ZWave.

## OMA - Open Mobile Alliance (http://www.openmobilealliance.org)

OMA provides a focal point for the development of mobile service enabler specifications. Today, it is a proponent of using the Light Weight Machine-to-Machine (LWM2M) protocol in IoT projects. This protocol can enable device management over sensor and cellular networks and can be used to transfer service data from a network to the devices. It also specifies device management protocols for mobile devices, service access, and connected IoT devices through OMA-DM.

## OMG - Object Management Group & Industrial Internet Consortium (http://www.omg.org)

OMG is the home of the Industrial Internet Consortium (IIC), a group of vendors creating industry use cases and test beds, reference architectures and best practices, and influencing global standards. Among the vendors taking part in IIC are AT&T, Cisco, GE, and IBM.

## OSGi Alliance (http://www.osgi.org)

The OSGi Alliance was created in 1999 to create open Java specifications. Today, it is helping to promote the usage of Java in IoT applications and gathers IoT demos and other proof points.

## Thread Group (http://www.threadgroup.org)

The Thread Group was formed in 2014 to create an IPv6 networking protocol designed for low power 802.15.4 mesh networks. Features include enabling mesh networks to be self-healing where hundreds of devices are deployed and secure encryption of data.

## TM Forum (http://inform.tmforum.org)

The TM Forum is a global industry association focused on providing research and publications targeting primarily service providers and technology suppliers. Its IoT focus area documents a wide breadth of current use cases and industry trends.

## Zigbee Alliance (http://www.zigbee.org)

The Zigbee Alliance now numbers over 400 members in its association. Its wireless solution sets are built using "Smart Objects" and are most often found in smart home products, connected lighting applications, and in the utilities industry (for monitoring, controlling, and automating the delivery of and usage of energy and water).

## Z-Wave Alliance (http://www.z-wavealliance.org)

The Z-Wave Alliance promotes the usage of Z-Wave technology for wireless control and monitoring of IoT devices and interoperability among Z-Wave devices. It also offers collaboration processes useful in development of new products and services. The products and applications produced by members of the alliance primarily focus on usage in the home and for light commercial activities.

# Index

# Get the eBook for only $5!

Why limit yourself?

Now you can take the weightless companion with you wherever you go and access your content on your PC, phone, tablet, or reader.

Since you've purchased this print book, we're happy to offer you the eBook in all 3 formats for just $5.

Convenient and fully searchable, the PDF version enables you to easily find and copy code—or perform examples by quickly toggling between instructions and applications. The MOBI format is ideal for your Kindle, while the ePUB can be utilized on a variety of mobile devices.

To learn more, go to https://www.apress.com/index.php/companion or contact support@apress.com.

CPSIA information can be obtained
at www.ICGtesting.com
Printed in the USA
LVOW04s0028250516
489756LV00002BA/19/P